# The Poetics of Angling in Early Modern England



Myra E. Wright takes ecocritical studies on an interdisciplinary turn toward the water with her new research monograph, *The Poetics of Angling in Early Modern England*. Identifying the lively presence of both literal and metaphorical images of sportfishing in all kinds of early modern writing, this book identifies a deep sympathy between the art of angling and the art of writing, and examines the centrality of fish in early modern conceptions of humanity.

**Myra E. Wright** is a Lecturer in English at Bates College, Maine.

**Perspectives on the Non-Human in Literature and Culture**
Series Editor: Karen Raber, *University of Mississippi, USA*

Literary and cultural criticism has ventured into a brave new world in recent decades: posthumanism, ecocriticism, critical animal studies, Object-Oriented Ontology, the new vitalism, Actor-Network Theory, and other related approaches have transformed the critical environment, reinvigorating our encounters with familiar texts, and inviting us to take note of new or neglected ones. A vast array of non-human creatures, things, and forces are now emerging as important agents in their own right. Inspired by human concern for an ailing planet, ecocriticism has grappled with the question of how important works of art can be to the preservation of something we have traditionally called "nature." Yet literature's capacity to take us on unexpected journeys through the networks of affiliation and affinity we share with the earth on which we dwell—and without which we die—and to confront us with the drama of our common struggle to survive and thrive has not diminished in the face of what Lyn White Jr. called "our ecological crisis." Animals have crept, slithered, trotted, swum and flown away from their customary function as metaphors, emblems, or analogies; now they populate critical analysis in increasingly complex ways, always in tension with our sense of what it means to be "human," or whether it is even possible to claim such a category of life exists. Posthumanism, which often shares terrain with ecocriticism and animal studies, has further complicated our conception of the cosmos by dethroning the individual subject and dismantling the comfortable categories through which we have interpreted our existence: we find our "selves" colonized by microbial beings, occupied and enhanced by mechanical inventions and contraptions, traversed by invisible forces from atmospheric changes to radiation, made vulnerable to our planet's sufferings, and inspired by its capacity for renewal. New materialism, new vitalism and all the permutations of theory that have evolved and are evolving to account for our entanglements with lively matter have ensured, however, that our displacement from the center of creation feels less like a demotion than a happy event: from our previous occupation of the impoverished, isolated space atop the mountain of creation, we might instead descend into the fecund plenitude of kinship with all things.

Until now, however, the elements that compose this wave of scholarship on non-human entities have had no place to gather, no home that would nurture them as a collective project. "Perspectives on the Non-Human in Literature and Culture" provides that structure, giving critical treatments of all kinds of non-humans and humans a local habitation. In this series, readers and fellow critics will find animals of all descriptions, but also every other form of biological life; they will meet the non-biological, the microscopic, the ethereal, the intangible. It is our goal for the series to provide an encounter zone where all forms of human engagement with the non-human in all periods and national literatures can be explored, and where the discoveries that result can speak to one another, as well as to readers and students.

**Birds and Other Creatures in Renaissance Literature**
Shakespeare, Descartes, and Animal Studies
*Rebecca Ann Bach*

**Race Matters, Animal Matters**
Fugitive Humanism in African America, 1838–1934
*Lindgren Johnson*

**Plants in Contemporary Poetry**
Ecocriticism and the Botanical Imagination
*John Charles Ryan*

**Meteorology and Physiology in Early Modern Culture**
Earthquakes, Human Identity, and Textual Representation
*Rebecca Totaro*

**Novel Creatures**
Animal Life and the New Millennium
*Hilary Thompson*

**Bestial Oblivion**
War, Humanism, and Ecology in Early Modern England
*Benjamin Bertram*

**Stages of Transmutation**
Science Fiction, Biology, and Environmental Posthumanism
*Tom Idema*

**The Poetics of Angling in Early Modern England**
*Myra E. Wright*

**Victorians and Their Animals**
Beast on a Leash
*Brenda Ayres*

# The Poetics of Angling in Early Modern England

Myra E. Wright

Routledge
Taylor & Francis Group
New York London

First published 2019
by Routledge
52 Vanderbilt Avenue, New York, NY 10017

and by Routledge
2 Park Square, Milton Park, Abingdon, Oxon OX14 4RN

First issued in paperback 2020

*Routledge is an imprint of the Taylor & Francis Group, an informa business*



*Library of Congress Cataloging in Publication Data*
A catalog record for this book has been requested

ISBN 13: 978-0-367-66493-0 (pbk)
ISBN 13: 978-1-138-30460-4 (hbk)

Typeset in Sabon
by Out of House Publishing

To the lakes and rivers of my youth, and all the creatures who splash, swim and drift in those cold waters

# Contents

| | | |
|---|---|---|
| | *List of Figures* | x |
| | *Acknowledgments* | xi |
| | Introduction: Facing Fish | 1 |
| 1 | Home Ecologies in the *Treatyse of fysshynge wyth an Angle* | 24 |
| 2 | Donne's Fish Biographies | 49 |
| 3 | The Fishing Lines of John Dennys | 75 |
| 4 | Shakespeare's Angling Devices in *Antony and Cleopatra* | 96 |
| 5 | The Interrupted Aquatic Hunt in Wroth's *Urania* | 116 |
| 6 | The Quickening of Walton's Incomplete Angler | 134 |
| | Conclusion: Killing and Conservation | 156 |
| | *Index* | 165 |

# Figures


| | | |
|---|---|---|
| 0.1 | The barbel (*Barbo barbo*) as depicted in Guillaume Rondelet's *Libri de piscibus marinis*, 1555 | 6 |
| 0.2 | The head of a "female salmon" in Pierre Belon's *De aquatilibus*, 1555 edition | 7 |
| 0.3 | A trout from Book 4 of Gessner's *Historiae animalium*, 1604 edition | 7 |
| 1.1 | Fish hooks from the *Treatyse of fysshynge wyth an Angle*, 1496 | 32 |
| 1.2 | The frontispiece from the *Treatyse of fysshynge wyth an Angle*, 1496 | 33 |
| 2.1 | Baited devices for capturing sea-pies, from Leonard Mascall's *Booke of fishing with Hooke & Line*, 1590 | 67 |
| 3.1 | The title page of John Dennys's fishing treatise, 1613 | 77 |
| 4.1 | Detail from the frontispiece of the *Treatyse of fysshynge wyth an Angle* | 106 |
| 4.2 | Detail from the frontispiece of *The Secrets of Angling* | 107 |
| 4.3 | Detail from the frontispiece of *The Compleat Angler* | 108 |
| 6.1 | A trout on offer in *The Compleat Angler*, 1653 | 135 |
| 6.2 | The frontispiece of the first edition of *The Compleat Angler*, 1653 | 138 |

# Acknowledgments

Many people have helped me bring this book to completion. Vin Nardizzi offered essential guidance at the earliest stages. Karen Raber, as series editor, has shown her typical generosity and enthusiasm. Jennifer Abbott, at Taylor & Francis, has been a patient editor and guide as I've worked on this first book. The anonymous readers who reviewed samples of the project provided invaluable ideas for improvement. I thank Isabel Stein for her patience and thoroughness as copy editor. My colleagues at Queens College, CUNY, gave their time and attention to a portion of this work in a faculty seminar. I'm especially grateful for the feedback of Steven Kruger, Glenn Burger, Hugh English, Gloria Fisk, Ryan Black, Richard McCoy, and Andrea Walkden. Three fellow teachers at Queens offered the kind of support that can come only from dear friends: Nicole Zeftel, Dara Barnat, and Meghan Fox. Christi Spain-Savage and Alena Buis have made me a better scholar with their good humor and eagerness to collaborate.

Over the past few years, meetings of the Shakespeare Association of America (SAA) have presented opportunities for me to share my work in seminars. I thank the gracious and thoughtful organizers of these: Miriam Jacobson and Wendy Hyman (Words and Things), Patricia Badir (Shakespearean Horizons), Vin Nardizzi (Histories of Sustainability), Suparna Roychoudhury and Mary Crane (Cognition in the Early Modern Period), and Matthew Hunter and Sam Fallon (Early Modern Cultures of Taste). I'm also indebted to my fellow seminar participants who have responded directly to parts of this project—especially Nicholas Helms, Douglas Clark, Amanda Lehr, Simon Smith, Anna-Claire Simpson, Jeffrey Theis, and Rob Wakeman. I'm grateful to the SAA for the contingent faculty grant that funded my attendance in 2016, and I'd like to thank the organization as a whole for the nursery of ideas it provides every year.

Ron Huebert and the faculty and students of Dalhousie University showed me a warm welcome in November 2016, when I traveled to share some of my work on Donne with the English department there. I thank Shannon Gayk for including me in a panel on Nature versus Ecology at the 2017 International Congress on Medieval Studies. I'm grateful to my fellow panelist Emily Houlik-Ritchey for her inspiring essay and the help she later gave me on the subject of neighbor-love. The brilliant staff of the Beinecke, Houghton, and Morgan libraries have brought treasures before my eyes. I thank these institutions and the talented people who share their resources with visitors like me.

Members of the faculty at Bates College, in my own department and beyond, have shown tremendous support for my teaching and research. I'm particularly grateful for conversations with Steve Dillon, Sylvia Federico, and Jane Costlow. The Faculty Scholarship Committee provided me with the development grant that has funded my research for the past year, and the Dean of the Faculty has covered my travel to conferences. I owe the world to my students—especially those who have taken my course The Lives of Rivers—for their hard work and their spirit of intellectual adventure. Nothing could be more inspiring.

And then there's family. I simply couldn't have done this work without the daily conversation and encouragement of Matt Frassica, who has never stood between me and the water. My parents have been behind me the whole way, and my brother Harlan was the project's first inspiration. A thousand thanks.

# Introduction
## Facing Fish

In an "admonition" to young women in love, published in 1567, the poet Isabella Whitney delivers conventional advice about how to manage a crush: "Trust not a man at the fyrst sight / but trye him well before" (A6v).[1] Using aquatic language throughout, the speaker introduces several watery images—dangerous mermaids, crocodile tears, and Leander's long-distance swim—before arriving at two fish whose paired narratives round out the lesson. The first of these is "carelesse," and misjudges his luck when he sees some food ahead: "He thinks his hap right good to bee / that he the same could spie, / And so the simple foole doth trust / to much before he trie" (A7v, A8r). Delighted by his own visual acuity but unable to see the hidden hook, this fish suffers the "spitefull fate" of being drawn from his habitat by an alien body. He is removed from his "happy state" in the water by the "cruel hands" of an unseen angler (A8r), and his life ends tragically on the shore. The speaker addresses this dead fish directly, telling him that he would still be enjoying the company of his friends and the warmth of the sunshine, if only he had seen the hook: "Then hadst thou with thy prety mates / bin playing in the streames / Whereas syr Phebus dayly doth / shew forth his golden beames" (A8r). The second fish, in contrast to his "felowe," has learned from a close encounter with the angler's gear that he must remain vigilant. Having once bit, he is twice shy: "he from fishers hooke did sprit / before he could him take, / And now he pries on euery baite, / suspecting styll that pricke: / (for to lye hid in euery thing) / wherewith the fishers stricke" (A8v).[2] Readers are directed to follow the second fish's good example, to "take hede," because even "the fish that reason lacks / once warned doth beware" (A8v).

In the final stanza, the speaker turns from the surprisingly astute fish to her own hard-won knowledge: "I who was deceiued late / by ones vnfaithfull teares / Trust now for to be ware" (A8v). Although the fish may lack reason—a contention that the underwater tableau seems to belie—it is the speaker's best way to figure forth the operations of her own lovelorn mind. Having survived the temptations of multiple hooks, this woman can address her own "prety mates" with confidence and jocularity. Because she has "sprit" from the dangers of someone else's sport, the speaker is at leisure to play a game of her choosing; she turns the predator's tackle into a collection of poetic devices that will protect her audience from deceit. The "yong Gentilwomen" reading Whitney's poem will receive information visually and change their behavior based on that information, just as the second fish does.

The poem's emphasis on eyes—from the fake tears a man might use to garner a woman's trust, to the imperfect vision of the first fish—invites comparisons between modes of observation and pursuit across species. If the fictional and heavily interpreted fish behavior in the poem strikes the modern reader as careless anthropomorphizing, a failure to behold the actions of other species without thinking first of human tendencies, it is nevertheless true that this speaker imagines herself as a fish. More precisely, she aims to tell us how fish see and think. In taking up this fantastical endeavor, Whitney's savvy advisor exhibits a typical early modern fascination with the cognitive faculties of aquatic species. She is certainly not the only one who looks fish in the eye.

Whitney's cautionary tale would make little sense in a culture where no one goes fishing with a rod and line—where, because there is no angling, there are no commonly held figures of thought that compare the human experience of being deceived to the piscine experience of getting caught. A poetics of angling—a system of imaginative engagements with the sport, borne out in text, image, and action—can be found in many different kinds of cultural objects from early modern England: poems like Whitney's, the plays of Shakespeare, prose fiction, and, of course, fishing manuals. This study maps how people of the sixteenth and seventeenth centuries thought about the sport of freshwater fishing, and asks what the representation of angling meant for the sport itself and the creatures pursued by its practitioners. In the chapters ahead, piscatory concerns are found in popular verse, drama, and prose. Some of these works are treatises on fishing, oriented toward the description, promotion, and refinement of the sport. Such practical texts are read alongside works that would have little application in the field (or the stream)—plays by Shakespeare, romance by Mary Wroth, poetry by John Donne—but that nevertheless rely on a cultural understanding of sportfishing in order to shape fictional worlds that include anglers and their prey. As a way of breaking down the boundary that has traditionally separated the practical literature from canonical works of tested literary merit, I read all of these texts in a similar way. All partake of literary traditions through their experiments in form and genre; all depend on the audience's curiosity and imagination. Further, each of the works examined here is situated in a changing ecosystem that includes not only fish and people, not only rivers and their banks, but the human habits of writing, printing, and reading. To present a clearer view of how literary practices might have influenced biotic environments (and vice versa), this study begins with the fish that lived in the freshwater systems of early modern England. Let us consider these creatures, face to face.

## Fish Faces, Fish Minds

Surely fish have faces—they have eyes, noses, and mouths that are not altogether different from our own, and are similarly backed by brains—but our tendency to picture fish in profile, with just one eye showing at a time, marks a failure to face fish, in two senses: fish are not confronted as other, more familiar (more mammalian) species are; and, as a consequence, they are not often recognized as distinct, individual beings. In fact, as Elspeth Graham points out, fish are only partially admitted into our broad company of living creatures: "They may be defined as animals (cold-blooded, gilled vertebrates) but they are not always, even in today's world, perceived as being quite animal."[3] To face fish is to acknowledge that some life forms are more difficult to know; they present a challenge to the imagination because of their apparently extreme bodily difference from us. But this is a difficulty that arises precisely because fish are not altogether alien. We know that their skin is slippery because we have held them, if only for a moment, in our own dry grasp.

In records of early modern smallholdings, Erica Fudge finds evidence "that in the first half of the 17th century recognizing that an animal had a face was central to owning and using that animal, and that—in a circular argument—the objectification of an animal actually emphasized its possession of a face."[4] While a human cannot share a home with a fish or stand staring into its eyes (both of which are possible in the case of the pigs, cows, and sheep described by Fudge), there is nevertheless a sense in which freshwater fish can be "knowable and known,"[5] if not individualized and named. The encounter between an angler and a caught fish is vanishingly brief in comparison with the long-standing domestic relationships found on a small family farm. Although both kinds of interaction might ultimately conclude with the killing of the non-human animal, the fish is never quite part of the household and is never expected to know its captor.[6] And yet it's the similarity between these situations that is most telling; in early modernity, fish, like mammalian livestock, were understood to have homes, families, distinct social roles, favorite foods, and all kinds of feelings. Fish had faces because they had minds.

Fudge's readings of household records have revealed that early modern people often attempted to interpret the appearance and behavior of animals; they looked for evidence of how a creature was feeling and what it was thinking. They identified even commodified animals as individuals. In short, they were capable of imagining that sheep, cows, and pigs had minds, and they used the idea of facial recognition to delineate a shared social world in which humans and other creatures might have similar cognitive experiences. The apprehension that animals could think about concepts like *home* did not necessarily lead, however, to relationships based on equality and mutuality. In Fudge's terms, "To possess a face does not simply mean that an animal becomes a member of the community, it allows for a level of complexity—symbolic, ethical, economic, religious—that is absent from the philosophies of Descartes, Milton, and Bacon, where animals are figured, in different ways, as simply bodies without minds."[7] The ironic compatibility of recognition and objectification that Fudge finds in early modern figurations of livestock is borne out even more strikingly in contemporary representations of piscine life. Early modern English fish are deadened, flattened, and removed from their environments in the very texts that celebrate their complex mental lives.

Recent studies of piscine cognition have shown that a fish's mind is hard to find. Although fish keep their brains behind their faces, like we do, their sensory faculties are distributed along their bodies and tucked away in surprising nooks and crannies. For instance, trout have chains of magnetite crystals in their noses, which seem to help them detect and measure geomagnetic fields.[8] Most fishes have an organ called the "lateral line"—a sensory canal that runs the length of the body and ends at multiple organs in the head. Sitting just beneath the skin, the lateral line has pores that lead to the fish's exterior.[9] Messages about water flow and turbulence travel back and forth between the brain and the lateral line. Among its many known functions and (we can assume) many that have not yet been discovered, the lateral line is responsible for telling a fish where it is, so that it can continue to shelter behind an obstacle in otherwise rapid water with little effort.[10]

As biologist Jonathan Balcombe has shown, science has moved away from the "cerebrocentric" perspective that dismissed fish cognition as "simple" on account of their relatively small brain size.[11] Instead, fish can show us how to revise our thinking about cognition. They are emblematic of the need to study "embodied cognition," to trouble the outmoded distinction between mind and body and accept the fact that the brain is just one part of a much larger cognitive story. Laurie Johnson, John Sutton, and Evelyn Tribble describe just such a shift in an introduction to a volume of essays on cognition in Shakespeare: "Increasingly new research in the cognitive sciences and the humanities alike no longer univocally supports privileging the active rational mind over passive biology .... Cognition is increasingly seen, within these new frameworks, as 'enactive,' 'embodied,' 'distributed,' 'situated,' or 'extended.' Emerging paradigms in the cognitive sciences have increasingly sought to extend cognition beyond the brain."[12]

As the brain is a relatively small piece of the world, there are many places for cognition to go. In Tribble's work alone, it travels into hands and feet, from actor to actor, around the physical space of the playhouse, and across the divide that separates so-called thought from so-called feeling. In her essay "Distributing Cognition in the Globe," Tribble draws on Edwin Hutchins's study of maritime navigation to argue that "human beings create smart structures predicated on a system of constraints which, paradoxically, enable the execution of complex tasks."[13] Thus, the "cognitive structure" at work in early modern theater can be seen to include "the playhouse, the plots, actors' roles, the plays' verbal structures, and the apprentice system and the organizational practices of the companies."[14] Investigating the development of particular skills in the "thinking body" of the early modern actor, Tribble contends, "A model of distributed cognitive ecologies is a particularly rich framework for studying skill, which links body, mind and environment in intelligent action." According to such a model, "the study of *skill* entails examining the training and the plasticity of the nervous system with habituation and practice; the role of attention, memory, and perception; and the extension of the body through a range of instruments and objects, including tools, treatises and social and material practices such as apprenticeship, enskilment, and enculturation."[15] Similarly, an investigation of early modern angling ought to position human bodies, rods, lines, bait, streams, and fishing manuals in one big system of informational exchange—not outside of the angler's brain, but connected to it in various dynamic ways.

Tribble's interventions can also help us conceptualize more broadly the cognitive operations of fish, without relegating their choices to the realm of irrational feeling. In fish as in humans, cognition and affect are intermingled. Tribble explains:

> I retain the word "cognitive," to mean, simply, thinking—an activity that is always both embodied and extended into the world. Thinking, that is, does not happen only in the head. Indeed, much current research supports the fundamental entanglement of processes that we label cognitive/mental or affective/bodily, and a more expansive view of the range of approaches across disciplines in the humanities and the sciences can provide a fuller account of the complexities of this relationship.[16]

To accept the existence of such complexities is to be better aligned with sixteenth- and seventeenth-century attempts to understand the life histories of fish. Without representing the sensory functions of the lateral line, the size of a fish's brain, or the magnetite crystals stashed in the nasal capsule of a trout, early modern English writers clearly demonstrate an interest in piscine behavior, and they take it for granted that fish exhibit intention, affection, hesitation, and frustration, among many other feelings.

Early modern fish, with their minds and faces, had been catalogued in several massive works of natural philosophy: the aquatic volume of Swiss naturalist Konrad Gessner's *Historiae animalium* (Book Four, 1558), and its major sources, both from France—Pierre Belon's *De aquatilibus* (1553), and Guillaume Rondelet's treatise in two parts, *Libri di piscibus marinis* (1554) and *Universae aquatilium historiae pars altera* (1555). These books present what looks at first like an affirmation of the mind–body divide—images of dead bodies alongside textual descriptions of lively, deliberate behavior—and yet to behold the faces of fish in each of these volumes is to ponder the limits of human vision and human sympathy. With their eyes wide open, the fish of Belon, Rondelet, and Gessner invite speculation about the thoughts and feelings of aquatic creatures.

The barbel pictured in Guillaume Rondelet's treatise has its characteristic beard-like strands of flesh and a suspicious sidelong glance (see Figure 0.1).[17] The face of the supposedly female salmon is so important in Pierre Belon's history of fishes that only her head appears—"Caput Salmonis foeminae" (Figure 0.2)—and at very close range.[18] The wide-open, toothy mouth and the stern striping above her eye suggest that this fish isn't here to make friends. The opposite might be assumed of Gessner's river trout, with its eager upward gaze and gently curved snout (Figure 0.3).[19] The precision of the woodcuts in all three treatises serves as evidence that the species have been closely observed, and that students of piscine life expect to learn how to differentiate among dozens of species based on visual cues. These detailed images were made possible by the *landing* of fish and the temporary convergence of human and piscine skin. Katherine Acheson presents an account of how such illustrations were meant to function:

> The conventions of realistic representation in these works were designed to counter the fantastic, over-determined, under-observed, animal forms in bestiaries, fable, and emblem collections, and to instantiate the practices and principles of scientific observation in the

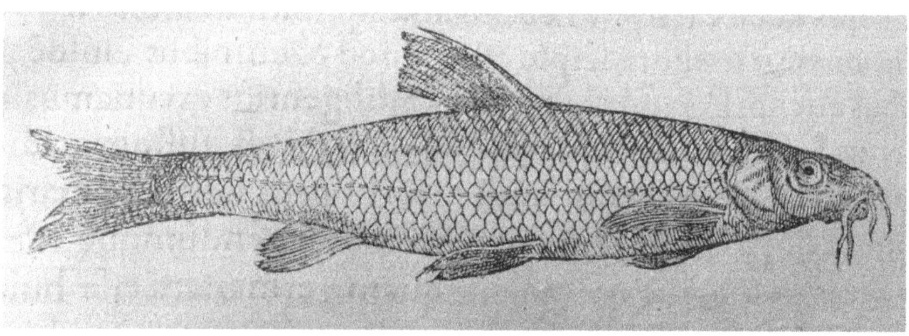

*Figure 0.1* The barbel (*Barbo barbo*) as depicted in Guillaume Rondelet's *Libri de piscibus marinis*, 1555. Beinecke Rare Book and Manuscript Library, Yale University.

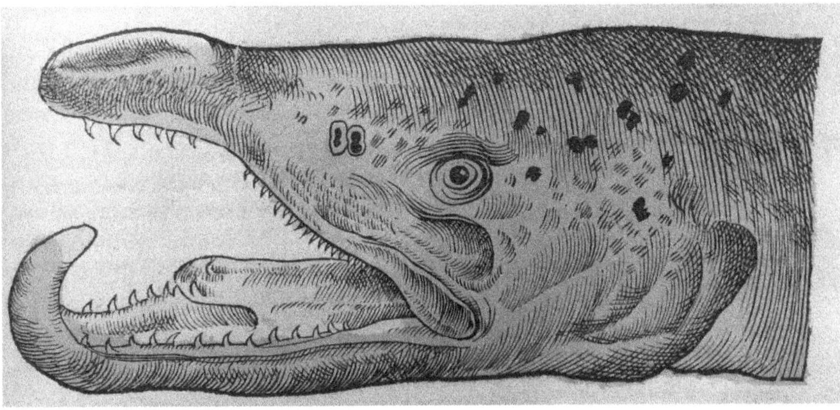

*Figure 0.2* The head of a "female salmon" in Pierre Belon's *De aquatilibus*, 1555 edition. Beinecke Rare Book and Manuscript Library, Yale University.

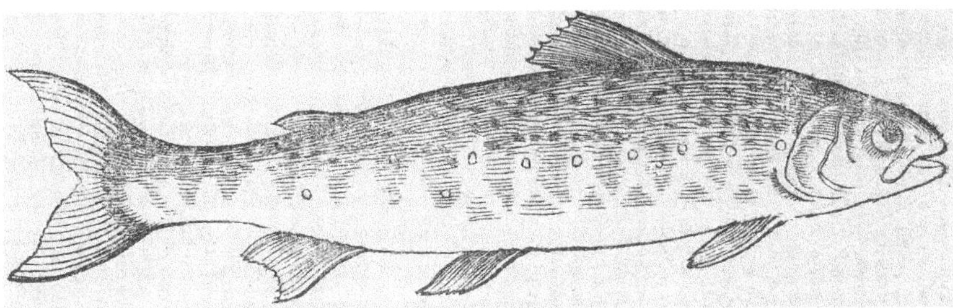

*Figure 0.3* A trout from Book 4 of Gessner's *Historiae animalium*, 1604 edition. Biodiversity Heritage Library. Digitized by Smithsonian Libraries.

> perception of the phenomenological world. These illustrations are intended to stand in for nothing further than the real animal itself; they were made from other drawings, or from dead specimens, and frame the creatures as objects within a collection.[20]

This assessment is confirmed by the remarkable degree of similarity between early modern treatises on fish and twenty-first-century field guides. In both genres, species are represented in such a way that they can be distinguished from one another in the wild. More importantly, the primary source of information in both cases is the activity of fishing, which brings these species into view. In the case of freshwater fishes, Gessner and his sources owe the precision of their images to angling above all else.

## Textual Habitats

The accuracy and relevance of the piscine images in Rondelet, Belon, and Gessner are by no means typical of sixteenth-century animal illustration. Chronicling a shift in the role of natural history, Marjorie Swann marks a clear break from medieval representations of life forms: "The seventeenth-century emphasis on individuals' experience of nature departed drastically from early concepts of what constituted knowledge of the natural world."[21] Before that shift, plants and animals were described with information found not in the field but in the library. For Swann, Gessner is representative of the earlier mode: "Full of myths and symbolism, works like Konrad Gessner's mid-sixteenth-century *Historia Animalium* exemplify what William B. Ashworth terms 'the emblematic world view' characteristic of much natural history written before the middle of the seventeenth century."[22] Thus we find in the ichthyological fourth book of Gessner a section devoted to mermen ("*De hominibus marinis*"), a lavish illustration of a hydra, and many other unlikely creatures. In her work on the origins of Gessner's images, Sachiko Kusukawa argues that the naturalist was concerned not only with the setting-down of observed truths:

> Gessner's *Historia animalium* was … an 'inventory' of knowledge about animals throughout history. Gessner did not distil or reduce similar descriptions, but rather juxtaposed them; nor did he eliminate contradictory or false descriptions of existing animals, or omit descriptions of animals whose existence was uncertain. The *Historia animalium* thus included familiar species such as horses, cats, dogs and mice, mythical beasts like the unicorn and the griffin, exotic animals including the crocodile, the sloth and the turkey, as well as various 'monsters'. Rarely was direct experience or observation the sole criterion for including an animal, while the existence of a textual description was.[23]

Kusukawa demonstrates that Gessner included both "true" and "false" images of creatures, and attached philological explanations that could help readers distinguish between a received mythology and a newly recorded observation from one of his many correspondents.[24] This labor of discrimination in itself represents a movement toward empiricism. Barbara J. Shapiro notes that "the term 'fact' or 'matter of fact' implied, for most of the seventeenth century, not something already worthy of belief or true but rather a matter capable of proof, preferably by multiple eyewitness testimony."[25] Poring over the images and text of Gessner's fourth volume, an early modern reader would not necessarily passively absorb information equally from all descriptions, but could arrive at the facts by scrutinizing the available sources collected therein and comparing the assertions of past authors to the experience of seeing living forms in nature.

As Swann shows, a change in the perceived function of natural history would draw people away from the archives and out of doors:

> Gradually ... a philological understanding of the natural world, rooted in knowledge of ancient texts, lost its monopoly over Europeans' apprehension of nature, and a new emphasis on "facts"—the products of firsthand observation—began to shape concepts of the role and practice of natural history within early modern culture.[26]

But when it comes to freshwater fish, we don't have to wait until the middle of the seventeenth century for accurate, non-symbolic representations of life forms. With their real-world accuracy, fish illustrations stand apart from many of the other images that constitute early modern treatises on natural philosophy. Many of the creatures are strange, but the pages upon pages devoted to freshwater fish in Belon, Rondelet, and Gessner include nothing that is so fantastical as to be omitted from modern field guides. Portraits and reports of various sea monsters appear in separate sections, and are treated differently from the piscine species that are said to inhabit European waters. Ultimately, these treatises demonstrate a desire to situate actual fish in their wild environments. In that respect, they were in the vanguard of a new way of thinking about why and how we ought to look at other creatures. Even the philological elements of the sixteenth-century ichthyologies, which connect observations across human cultures, can be understood as advancements rather than regressions.

In all of the natural histories of fish from the middle of the sixteenth century there is a striking emphasis on geography. Both Belon and Rondelet organize their treatises according to habitats, separating freshwater species from the larger catalogues of marine fish and marking off separate sections for the fishes of lakes and rivers. In compounding their works with the findings of others, Gessner does away with Belon and Rondelet's designation of habitats, mixing saltwater and freshwater species in a comprehensive, alphabetized survey. Under the broad heading of "whitefish," though, Gessner places an unusual emphasis on local geography by including a subsection on the fish of Lake Leman, listing species of whitefish before veering off course to include various other kinds of fish found in that particular body of water. He borrows the designation "De alio pisce proprio Lemani lacus" from Rondelet.[27] The chorographic tendency of these treatises is borne out also in multilingual taxonomic information—each species is named not once but many times, so that readers can learn what a pike might be called in Greek, Latin, and several European vernaculars, all at once. In the tables of contents that preface

Gessner's text, handy lists of species names appear in many languages. The philological apparatus of these works gestures backwards to historical precedents for scientific observation. It also acknowledges the fact that similar species can be found in places that bear distinct human cultures; implicitly, the variety of names for each creature serves as a reminder that piscine populations and human populations have different kinds of mobility and diversity. As Elspeth Graham puts it:

> Fish are unfixed animals: although different fish are specific to the particularities of geographic region and local conditions (char in Lancashire; the large roaches of the Thames; the trout of Hampshire; herring, cod, and eels from eastern English seas, for instance), they are, by their nature, mobile in the environment. They can be metonyms of place while always transcending place.[28]

Fish do not inhabit the political world in quite the same way that people do. They can move from place to place and be called by several different names without losing their particular creaturely identity. But certain fish do *belong* in certain places: the salmon of Lake Leman, for example; or the grayling of the River Lambro.[29] These are fish with homes. In treatises that are packed with portraits of alien creatures, it is interesting to find that some are more foreign than others. The reader is encouraged to choose an orientation relative to the species being described. Writing in Latin for Swiss publication, Gessner at one point cites his correspondence with the prominent English physician William Turner. In so doing, he draws attention to the lives of foreign fishes: "The English call the Trout *Trutte* or *Trouut*. There are two kinds of trout found in England (as the learned man William Turner has written to me)."[30] Readers can perform a repetition of Gessner's research by flipping back to a transcription of the letter that contains this information, printed as a prefatory epistle to the treatise. The philological apparatus, in all its glorious redundancy, invites an important question: Is there such a thing as an English fish?

This question becomes rather poignant when we consider that Gessner's encyclopedic text came into the English language with its fish removed. Despite its central reliance on the *Historiae animalium* as a source text, Edward Topsell's monumental *Historie of Foure-footed Beastes* and *Historie of Serpents* (1607–1608) shows us a world without fishes. The sensitive depictions of fish from sixteenth-century texts come down to posterity not through Topsell, but through Izaak Walton, who pairs them with advice about how to catch each species. Since the first publication of *The Compleat Angler* in 1653, the categorization and illustration of fish has been a practice closely associated with angling. It seems that piscatorial literature came to fill a gap in natural philosophy by providing images and accounts of creatures who were not otherwise being described in a systematic way. Another consequence of Topsell's failure to bring Gessner's collection of piscine images into an anglicized text is that the portrayal of fish in early modern England was, until the middle of the seventeenth century, a task carried out predominantly in words rather than pictures. It may be that the textual descriptions of piscine life from the sixteenth and early seventeenth centuries owe some of their depth, richness, and variety to a relative dearth of fish illustration in the period.

The readings that make up this book include precise textual renderings of fish, sometimes of particular species. Could early modern people picture such creatures as they read a line of poetry or heard a character speak onstage? It is possible that the average person had much more ichthyological knowledge than most of us do in this age of scientific inquiry. The detailed portraits offered up in the text of English angling treatises, like the continental natural histories of fish that preceded them, address an audience that is familiar with trout, eels, and pike, and eager to learn more about these neighboring animals. When early modern writers depict piscine life, whether their task is to deliver practical instruction or potent metaphor, they face fish.

## A Change of Scenery

Since 2009, ecocritical work on early modernity has turned toward the water, but this water is usually the briny kind—Steve Mentz and Dan Brayton have urged us to explore Shakespeare's seas, and Elizabeth J. Bellamy has argued compellingly for the importance of England's coastlines in the literature of the sixteenth and seventeenth centuries.[31] These studies, along with many others, have brought marine habitats into focus for readers of early modern literature. And as ecotheorists move "beyond green,"[32] the standard emphasis on terrestrial verdancy gives ground not only to the big blue sea, but also to the less azure waters of rivers, lakes, and ponds. Angling literature exhibits a range of hues, including a very different kind of green: the emerald of deep, clear lakes. But fresh waters are not always clear. Fishers consistently emphasize the problems of visual apprehension as they reckon with depths that can both magnify and distort their contents. The medium through which an angler gazes is not a transparent filter, but a dynamic system of light that involves quantities of vital and inanimate matter. Thus, biologists who study aquatic environments are attuned to the meanings of color: "Generally, in natural waters where there are small amounts of suspended matter, the water looks green, but where there are large amounts, the water looks yellow or brown."[33]

This prismatic range indicates just how varied bodies of fresh water are; in setting saltless waters apart from seas, we have established a massively diverse category of geographical features, from torrents and pools to bogs and fens. To study the history of angling is to consider not only the limpid greens and blues of rushing mountain streams and placid lakes, but also the many browns that might be found in a healthy fish habitat. Declaring that brown "is no one's favorite color," Steve Mentz lays out its importance for our understanding of biotic environments:

> A color you cannot see through, brown captures a connecting opacity at the heart of ecological thinking. It comes at us from both sides of our world, the living and the dead. Brown marks the fertile soil that plants consume and the fecal waste that animals reject. Thinking brown pushes us into hybrid spaces that span living and nonliving matter, aesthetic values and biological drives.[34]

Anglers are surely the exception that proves Mentz's rule, with their preference for drab attire, their intimate familiarity with mud and silt, their procurement of pheasant hackle for fly-tying, and their pursuit of the brown trout (*Salmo trutta*). John Dennys writes in his fishing treatise, published in 1613, about the angler's need to disappear into a brown environment: "let you[r] garments Russet be or gray, / Of colour darke, and hardest to discry; / That with the Raine or weather will away, / And least offend the fearefull Fishes eye."[35] Furthermore, in all early modern angling manuals there is a demonstrated interest in shit, the quintessential brown substance. Manure is not only the rejected matter of one species, but the welcome habitat of another; anglers who would seek the best live bait are instructed to harvest fly larvae from cow dung.

The confluence of life forms and nonliving material that Mentz locates in *brown* is equally characteristic of *slime*, as it is rendered by Dan Brayton. Slime is "a mediating term between living and inanimate objects."[36] It also happens to be the defining feature of a fish's touch, and in that sense "a mediating term" between human and piscine bodies. As Elspeth Graham observes, "Fish are creatures who slip away."[37] Their slimy exteriors protect them from many eventualities, one of which is capture by humans and other land-dwelling predators. And there are other mucilaginous phenomena in the angler's environment: the glistening goop between the eggs in a roe sac, the slick on a dangled worm, the algae that grow on rocks by the river, and the glossy coating on semi-aquatic plants. Angling and its literatures take place in a world that accommodates shadows, puddles, clouds, mud, and slime aplenty. The wildflowers, formidable trees, and sweet breezes of early modern romance are all found in contemporary descriptions of sportfishing, but anglers also delight in things that others would rather avoid. To write about freshwater fishing—in the seventeenth century or our own—is to ascribe a recreational value to cold weather, isolation, and insects. Angling takes us where we might not otherwise go.

Moreover, the sport depends upon an understanding of water systems and the larger geographical schemes into which they fit. Freshwater systems can be divided into two basic categories: standing waters and running waters.[38] With very few exceptions, these two kinds of system are usually somehow connected to each other and ultimately to the sea. It is clear from the natural philosophy of the sixteenth and seventeenth centuries that early modern Europeans distinguished between fresh and salt waters. As many readers of the period's literature have shown, the most important bodies of fresh water for the English were rivers. In her recent study of shorelines, Elizabeth J. Bellamy acknowledges the cultural centrality of rivers: "If asked to think about bodies of water in early modern England, what most readily comes to mind are … rivers, the dominant principle in early modern English topography."[39] Maggie Kilgour writes about the political significance of rivers for early modern Britain, an island nation shot through with waterways: "Drawing upon Christian and neoplatonic visions of the many flowing into a single source, the image of the rivers of a Britain surrounded by a sea suggested a national independence and wholeness that was achieved through the harmonious reconciliation of internal differences."[40] Rivers and their edges dominate not only the chorographic projects of Tudor river poetry, but also the emerging genre of angling literature, which privileges riparian zones above all other geographical locations.

Fish biologists consider the distinctions among various water systems when they describe a piscine population as anadromous, catadromous, or non-migratory. Anadromous fish start their lives in fresh water and return to spawn after a sojourn in salt water. Catadromous fish, like eels, do the opposite. Non-migratory species, including landlocked populations of typically migratory species, stay put in their home waterway for a lifetime. Where fish choose to live has to do with climate change and evolution. About fifteen thousand years ago, during the last great Ice Age, fish living in the part of the world we now call Europe could survive only by huddling in a few refuges. When the ice began to melt, species moved out of the refuges into habitats that were accessible and hospitable. With a sea level much lower than today's, there was no English Channel, so even fish that could not tolerate salt water could migrate from Germany to southeast Britain through freshwater systems.[41] But geology often prevented colonization; fish in need of fresh water could not migrate up high waterfalls or across mountain chains that had no rivers.[42] Eventually, the seas filled up and kept salt-intolerant species from travelling. The result of all this climate change was that some species of freshwater fish were distributed all over Europe, while others were drastically localized. Some fish, it seems, are more English than others.

Furthermore, the behavior of anadromous and catadromous species reminds us that fresh water runs to the sea and mixes with it, disturbing the very notion of a marked distinction between freshwater and saltwater creatures. Although Atlantic salmon (*Salmo salar*) have been known to hesitate as they enter salt water during the smolting phase of their life cycle, their body systems have developed in such a way that they can automatically balance their internal levels of salt and water in order to survive in either a stream or an ocean, without pausing to figure it out.[43] Fish may not think about hydrology as humans do, but they reckon with the same phenomena of currents, confluences, and pollutants that we seek to measure in their habitats. In fact, humans get information about waters from the apparent decisions of fish. Ecologists Peter Maitland and Neville Morgan describe research that relies on the authority of piscine behavior:

> Over the last two decades, many freshwater and estuarine species have returned to the lower Thames, which now supports a diverse community not unlike its original one. Rehabilitation of the River Clyde has been rather slower. However, salmonid fish are conclusive evidence of high water quality and the return of the Atlantic salmon to this river after an absence of more than 100 years is a tribute to decades of work by the local river purification board.[44]

Early modern piscatory writing does not reflect a detailed understanding of the homing patterns of fish—something that researchers are still struggling to achieve today—but it does draw attention to the importance of particular environments for particular species. Like modern-day monitors of habitat rehabilitation on the River Clyde, angling enthusiasts of early modernity gathered information about environments by reading the actions of fish. Writing sometime before the end of the fifteenth century, the author of the *Treatyse of fysshynge wyth an Angle* was already interpreting the behavior of *Salmo salar* in relation to its home stream: "The samon is a gentyll fysshe: but he is comborous for to take. For comynly he is but in depe places of grete ryuers. And for the more parte he holdyth the myddes [the midst] of it: that a man may not come at hym."[45]

## A Cognitive Ecology of Angling

The chapters that follow look closely at such imaginative moments in early modern texts, when an angler is seen to speculate about the reasons for fish behavior. Each reading approaches angling as a complex element of the environment in which a fish might live. The sport is a very particular kind of interspecies interaction that has evolutionary repercussions for humans and fish alike. If we construe the sport more broadly, and accept its literal and figurative representations as part of its ecological profile, we can begin to develop a cognitive ecology of angling itself. A deep view of how sportfishing has shaped humans, fish, and habitats requires not only images of how angling is performed, but also descriptions of the imaginative functions the sport has served in culture over time. An ecology of angling necessarily finds the angler in an environment and measures his or her interactions with the surrounding phenomena. A *cognitive* ecology of the sport focuses on the operations of information processing that make any fishing scene—and any representation of such a scene, however figurative—a rich locus of scientific and cultural meaning.[46] Like recent studies that illuminate the inseparability of mind and body in early modern drama, and the shared, embodied, affective responses to stagecraft that might have been experienced by an early seventeenth-century audience at the Globe,[47] this project aims to understand environments by paying careful attention to the interactions among their inhabitants.

At the same time, many of the readings that make up this study require a dismissal of the commonly held belief in a clear dividing line between subjects and environments. Following Lawrence Buell, I have adopted a "myth of mutual constructionism: of physical environment (both natural and human-built) shaping in some measure the cultures that in some measure continually refashion it."[48] In a similar vein, Nancy Tuana introduces the term *interactionism* to accommodate a flexible, ongoing exchange between human selves and the worlds in which they live:

> Interactionism acknowledges the robust porosity between phenomena that destabilizes any effort to finalize a nature/culture divide. We can, and often need, to make distinctions between such poles, but it is crucial not to see these distinctions as "natural kinds" or to read them as reflecting a dualism .... Interactionism not only allows but compels us to speak of the biological aspects of phenomena without importing the mistaken notion that this biological component exists somehow independent of, or prior to, cultures and environments. It serves as witness to the materiality of the social and the agency of the natural.[49]

The materiality of the social, for my purposes, includes the phenomena that constitute literary production—not just written and printed texts or the practices of composing and reading them, but the networks of ideas that make texts intelligible within a culture. Commonly held assumptions, standard modes of representation, and ordinary turns of phrase are all phenomena that interact with other elements in our environment. We can think of such cultural habits, in fact, as "biological components" in their own right.

At times, the readings that make up this study adopt ideas from conceptual metaphor theory, as articulated by cognitive linguists like Mark Johnson and George Lakoff and cognitive psychologists like Raymond Gibbs. The present study places texts that clearly have a practical orientation toward an audience of actual anglers alongside works that share no instructional wisdom with readers but are invested in specific cultural meanings of the sport. Figurative deployments of angling were potent in early modern English culture because the sport itself was already, for its practitioners, a way of thinking through certain social and literary problems: how to persuade, how to deceive, how to choose a place in the world, how to interpret the behavior of others, how to wait. Proponents of conceptual metaphor theory believe that the comparative frameworks of commonplace metaphors grow out of lived human experience;[50] the idea of falling for something "hook, line, and sinker" is evocative precisely because of a perceived difference between human beings (predators) and fish (prey) that has been emphasized and exploited by the sport of angling.

Ideas from conceptual metaphor theory resonate with uses of cognitive ecology in literary studies, especially when readers try to assess how a widespread figure of speech has come to have a precise meaning for a broad range of people within a culture. Raymond Gibbs has argued that "cognitive linguists and cognitive psychologists ... should think about metaphor and its relation to thought as cognitive webs that extend beyond individual minds and are spread out into the cultural world."[51] If we combine the basic claims of conceptual metaphor theory with recent findings in cognitive ecology, it becomes possible to identify a complex relationship between the practices of anglers and the representation of the sport. I argue that depictions of angling, whether figurative or literal, have real ecological consequences that are hard to measure. The bounty of representations that made angling familiar and intelligible for early modern people—expressions that cultivated and confirmed conceptual metaphors like "to deceive someone is to catch a fish" or "an unfortunate person is a fish"—must be treated as real elements in living ecosystems, not as one-sided human interpretations of static environments. Early modern representations of angling can provide us with plenty of instruction on how to observe the transit of signals from one body to another. When poets, playwrights, and writers of romance frame piscatory metaphors or conjure images of anglers by the water, they are issuing a particular kind of signal that can still, after several centuries, be received by readers.

In Isabella Whitney's "admonition," the second fish "pries" on all passing bait, suspicious of the seeming treats in its vicinity. To "pry" is "to look, [especially] closely or curiously; to peer inquisitively or intrusively; to seek out secret or private information, to spy."[52] Cognitive ecologists study this kind of looking when they use signal detection theory to describe the feeding behavior of fish. As Thomas Grubb explains in his wonderful book *The Mind of the Trout*, there are four possible responses to a perception:

> First, when the signal really is there, the fish can display what is known as *correct detection* by swimming out, collecting the food item, and returning to its station. Second, the fish can fail to respond to a real object; this alternative is termed a *missed detection*. Third, the fish can make a mistake by swimming out after a perceived object

> that actually is not there at all, a response well termed a *false alarm*. Finally, the trout can correctly conclude that no signal exists and stay put. This last alternative is called a *correct rejection*.[53]

The first fish in Whitney's poem exhibits the "false alarm" kind of signal response; the second performs a "correct rejection." The matter of signal detection can be further complicated, as the "admonition" shows, by a central feature of the angler's sport: the production of deceptive signals. Like recent studies of signal detection, the poem's riparian scene explores the exchange of information between two organisms—a signaler and a receiver. Anglers, like many animals, benefit from producing false alarms in other creatures. The sportfisher's bait is a "parasitic signal" specifically designed to "mimic another signal (the model signal) that normally elicits a correct detection."[54] The rhetorical destination of Whitney's speaker is another kind of parasitic, deceptive signal: the "vnfaithfull teares" of a false lover, which have also been apprehended visually and interpreted with reference to a foregoing pattern of experience. Fish and gentlewoman alike must develop strategies of counterintelligence in order to survive. Both must learn to pry. And so must we, if we are to make the most of the signals issued within the system of figures upon which early modern authors relied.

The chapters that follow introduce texts in a loose chronology, starting with the earliest known angling treatise in English and ending with the most celebrated work in this tradition: Izaak Walton's *Compleat Angler*. Practical texts are intermingled with works of imaginative literature in order to maintain a rough timeline and—more importantly—to trouble the generic and disciplinary boundaries that would flatten a reading of any of these works. The first chapter presents the "home ecologies" of the anonymous *Treatyse of fysshynge wyth an Angle* (1496).[55] Here, I consider how the patch ecosystems described in the *Treatyse* are seen to accommodate a staggering range of neighboring phenomena—from God to maggots—even as the text itself strives to limit its readership for the good of the sport. The second chapter traces a fascination with the beginnings and ends of piscine life in the poetry and sermons of John Donne, focusing in particular on the larval fish and violent deaths that appear in his long poem *Metempsychosis*. In the following chapter I examine the formal features and didactic momentum of John Dennys's *Secrets of Angling*, which carries the distinction of being the first English fishing treatise in verse. I argue that the very structure of the poem is what makes it a useful resource for the hopeful angler. In the *Secrets*, angling is advanced as a vernacular sport that is particularly suited to the familiar networks of English rivers and the cognitive faculties of the humans and fish who inhabit those riparian zones.

The fourth chapter is a brief study of the only anglers to be found in Shakespeare's drama: Antony and Cleopatra. Because these characters and their chosen sport come down to the playwright from Plutarch through two translators, my reading of their recreation hinges on a philological treatment of angling in the play and its sources. In Mary Wroth's *Urania*, the subject of my fifth chapter, angling is often interrupted by the wandering thoughts of the fisher or the greater claims of social interaction. The *First Part* of the romance presents two scenes that begin as quiet piscatory resorts for individual anglers but quickly become sites of convergence and revelation, where strangers catch the attention of the fisher before she can catch any fish. The sixth chapter arrives at Walton's monumental treatise *The Compleat Angler* (1653). I chart the development of Walton's novice angler with reference to two available models of transformation, both of which are mentioned within the narrative frame: religious conversion and biological metamorphosis. Finally, the study concludes with a consideration of the enduring association between angling and practices of conservation. Although the killing of animals is an undeniable element of sportfishing, angling has been credited with the protection of habitats and species. Noting the long history of this paradox and the ethical problems attendant upon it, I ask what else we can learn from an early modern poetics of angling.

In quoting early editions of primary texts, I have retained all original spelling and punctuation, with these exceptions for ease of reading: long S is shortened, I and J are occasionally modernized, and superscripts are expanded.

## Notes

1 [Isabella Whitney], "The admonition by the Auctor, to all yong Gentilwomen: And to all other Maids being in Love," in *The copy of a letter, lately written in meeter, by a yonge gentilwoman: to her vnconstant louer* (1597, STC [2nd ed.] 25439). Text citations refer to signature numbers in this edition. A handy edition of the poem is available at *Representative Poetry Online*, ed. Ian Lancashire (University of Toronto Libraries, 2003). https://tspace.library.utoronto.ca/html/1807/4350/poem2952.html.

2 The verb "stricke" might simply be an alternative rendering of *strike*, or it may be a misprint for *tricke* (having doubled and carried the S from the end of "fishers"). (*OED*, s.v. "strike, v.," sense 25, accessed May 4, 2018) www.oed.com/view/Entry/191633?rskey=SlhFdo&result=3&isAdvanced=false

3 Elspeth Graham, "Ways of Being, Ways of Knowing: Fish, Fishing and Forms of Identity in Seventeenth-Century English Culture," in *Animals and Early Modern Identity*, ed. Pia F. Cuneo (Burlington, VT: Ashgate, 2014), 351.

4 Erica Fudge, "The Animal Face of Early Modern England," *Theory, Culture & Society* 30 (2013): 195.

5 Fudge, "The Animal Face," 185.

6 A further complication in the story of fish/human interaction is the fact that fish and the bait used to catch them were forms of livestock for some people in early modernity. Fish ponds were stocked, maintained, and fished by those

whose properties were capacious enough to accommodate them, and it would be possible for such a landowner to angle for the same fish, at home, more than once. As for bait, anyone with the time and enthusiasm to angle could gather live bait and learn how to tend it.

7 Fudge, "The Animal Face," 190.

8 Thomas C. Grubb, *The Mind of the Trout: A Cognitive Ecology for Biologists and Anglers* (Madison: University of Wisconsin Press, 2003), 32.

9 Peter S. Maitland, *The Hamlyn Guide to Freshwater Fishes of Britain and Europe* (London: Hamlyn, 1977), 10.

10 Grubb, *The Mind of the Trout*, 23.

11 Jonathan Balcombe, *What a Fish Knows* (New York: Scientific American / Farrar, Straus and Giroux, 2016), 12.

12 Laurie Johnson, John Sutton, and Evelyn Tribble, introduction to *Embodied Cognition and Shakespeare's Theatre* (New York: Routledge, 2014), 3.

13 Evelyn Tribble, "Distributing Cognition in the Globe," *Shakespeare Quarterly* 56, no. 2 (2005): 141.

14 Tribble, "Distributing Cognition," 142.

15 Evelyn Tribble, *Early Modern Actors and Shakespeare's Theatre: Thinking with the Body* (London: Bloomsbury Arden Shakespeare, 2017), 5.

16 Evelyn Tribble, "Affective Contagion on the Early Modern Stage," in *Affect Theory and Early Modern Texts*, ed. Amanda Bailey and Mario DiGangi, 200–201 (New York: Palgrave Macmillan [Springer Nature], 2017), DOI: 10.1057/978-1-137-56126-8.

17 Guillaume Rondelet, *Libri de piscibus marinis, in quibus veræ piscium effigies expressæ sunt*, vol. 2 (*Uniuersae aquatilium historiæ pars altera*) (Lyon, 1555), 194.

18 Pierre Belon, *De aquatilibus, libri duo* (Paris, 1553), 279. Incidentally, this image looks much more like a spawning male than a female, because of the tell-tale hook curving up from its lower jaw. The curvature is one of several skeletal changes that take place as salmon prepare to migrate upstream from the sea. Witten and Hall note: "The most prominent alteration of the head skeleton is the development of a kype (hook) at the distal tip of the lower jaw of maturing males. During the sea-life stage, males and females are morphologically similar.... Development of the kype during migration establishes, in part, the sexual dimorphism between males and females; female salmon do not develop a kype" (P. Eckhard Witten and Brian K. Hall, "Seasonal changes in the lower jaw skeleton in male Atlantic salmon [*Salmo salar* L.]: remodelling and regression of the kype after spawning," *Journal of Anatomy* 203 (2003): 435–36).

19 Konrad Gessner, *Historiae animalium, Liber IV, Qui est de Piscium & Aquatilium Animatium natura* (Frankfurt, 1604), 1007.

20 Katherine Acheson, *Visual Rhetoric and Early Modern English Literature* (Burlington, VT: Ashgate, 2013), 131.

21 Marjorie Swann, *Curiosities and Texts: The Culture of Collecting in Early Modern England* (Philadelphia: University of Pennsylvania Press, 2001), 57.

22 Swann, *Curiosities and Texts*, 57–58.

23 Sachiko Kusukawa, "The Sources of Gessner's Pictures for the *Historia animalium*," *Annals of Science* 67, no. 3 (July 2010): 306.

24 Kusukawa, "The Sources of Gessner's Pictures," 327.

25 Barbara J. Shapiro, *A Culture of Fact: England, 1550–1720* (Ithaca, NY: Cornell University Press, 2000), 110.
26 Swann, *Curiosities and Texts*, 59.
27 Gessner, *Historiae animalium,* 31; Rondelet, *Libri de piscibus marinis,* 164.
28 Graham, "Ways of Being," 353.
29 Rondelet, *Libri de piscibus marinis,* 160; Belon, *De aquatilibus,* 282.
30 Gessner, *Historiae animalium,* 1009.
31 Steve Mentz, *At the Bottom of Shakespeare's Ocean*, Shakespeare Now! (London: Continuum, 2009); Dan Brayton, *Shakespeare's Ocean: An Ecocritical Exploration*, Under the Sign of Nature (Charlottesville: University of Virginia Press, 2012); Elizabeth J. Bellamy, *Dire Straits: The Perils of Writing the Early Modern English Coastline from Leland to Milton* (Toronto: University of Toronto Press, 2013).
32 I borrow this suggestive phrase from the subtitle of *Prismatic Ecology: Ecotheory beyond Green*, a volume of essays edited by Jeffrey Jerome Cohen (Minneapolis: University of Minnesota Press, 2013).
33 Peter S. Maitland, *Biology of Fresh Waters* (New York: Wiley, 1978), 4.
34 Steve Mentz, "Brown," in *Prismatic Ecology*, 193.
35 [John Dennys], *The Secrets of Angling* (1613, STC [2nd ed.] 6611), B5v.
36 Dan Brayton, "Shakespeare and Slime: Notes on the Anthropocene," in *Ecological Approaches to Early Modern Texts: A Field Guide to Reading and Teaching*, ed. Jennifer Munroe, Edward J. Geisweidt, and Lynne Bruckner (Burlington, VT: Ashgate, 2015), 84.
37 Graham, "Ways of Being," 351.
38 Maitland, *Biology of Fresh Waters*, 62.
39 Bellamy, *Dire Straits*, 8.
40 Maggie Kilgour, "Writing on Water," *English Literary Renaissance* 29 (1999): 285.
41 Malcolm Greenhalgh, *The Pocket Guide to Freshwater Fish of Britain and Europe* (London: Mitchell Beazley, 2001), 9.
42 Greenhalgh, *The Pocket Guide*, 8.
43 Grubb, *The Mind of the Trout*, 36–38.
44 P. S. Maitland and N. C. Morgan, *Conservation Management of Freshwater Habitats* (London: Chapman & Hall, 1997), 161.
45 *This present boke shewyth the manere of of hawkynge [and] huntynge and also of diuysynge of cote armours…* [*The Book of Saint Albans*], 1496. STC (2nd ed.) 3309, h6r.
46 Reuven Dukas, *Cognitive Ecology* (Chicago: University of Chicago Press, 1998), 1.
47 I am thinking here of the essays that make up Johnson, Sutton, and Tribble's collection *Embodied Cognition and Shakespeare's Theatre* (Routledge, 2014).
48 Lawrence Buell, *Writing for an Endangered World: Literature, Culture, and Environment in the U.S. and Beyond* (Cambridge, MA: Harvard University Press, 2001), 6.
49 Nancy Tuana, "Viscous Porosity: Witnessing Katrina," in *Material Feminisms*, ed. Stacy Alaimo and Susan Hekman (Bloomington: Indiana University Press, 2008), 192, 209–10.
50 George Lakoff and Mark Johnson, *Metaphors We Live By* (Chicago: University of Chicago Press, 1980), 14.

51 Raymond W. Gibbs, Jr. "Taking Metaphor Out of Our Heads and Putting It Into the Cultural World," in *Metaphor in Cognitive Linguistics: Selected papers from the fifth International Cognitive Linguistics Conference, Amsterdam, 1997*, ed. Raymond W. Gibbs, Jr. and Gerard J. Steen (Amsterdam and Philadelphia: John Benjamins, 1999), 146.

52 *OED*, s.v. "pry, v.1" (September 2007). www.oed.com/view/Entry/153665?rskey=R24PNH&result=5&isAdvanced=false.

53 Grubb, *The Mind of the Trout*, 72.

54 Grubb, *The Mind of the Trout*, 78.

55 In a seminar entitled "Home Ecologies" at the 2017 meeting of the Shakespeare Association of America, Mary Trull and Jennifer Munroe invited participants to put the *oikos* back into *ecology*: "Drawing upon scholarship on early modern households and their material practices by authors including Wendy Wall and Natasha Korda, this seminar revisits our notion of 'ecology' by maintaining the emphasis on material bodies and objects but including how physical environments, living and non-living, human and non-human, inform our understanding of multiple exchanges in and related to the home/household" (seminar proposal, personal correspondence). I'm grateful to Trull and Munroe for letting me use their seminar title in this project.

## References

Acheson, Katherine. *Visual Rhetoric and Early Modern English Literature*. Burlington, VT: Ashgate, 2013.

Balcombe, Jonathan. *What a Fish Knows: The Inner Lives of Our Underwater Cousins*. New York: Scientific American / Farrar, Straus and Giroux, 2016.

Bellamy, Elizabeth Jane. *Dire Straits: The Perils of Writing the Early Modern English Coastline from Leland to Milton*. Toronto: University of Toronto Press, 2013.

Belon, Pierre. *De aquatilibus, libri duo*. Paris, 1553.

Brayton, Dan. "Shakespeare and Slime: Notes on the Anthropocene." In *Ecological Approaches to Early Modern Texts: A Field Guide to Reading and Teaching*, edited by Jennifer Munroe, Edward J. Geisweidt, and Lynne Bruckner, 81–90. Burlington, VT: Ashgate, 2015.

———. *Shakespeare's Ocean: An Ecocritical Exploration*. Under the Sign of Nature. Charlottesville: University of Virginia Press, 2012.

Buell, Lawrence. *Writing for an Endangered World: Literature, Culture, and Environment in the U.S. and Beyond*. Cambridge, MA: Harvard University Press, 2001.

Cohen, Jeffrey Jerome, ed. *Prismatic Ecology: Ecotheory Beyond Green*. Minneapolis: University of Minnesota Press, 2013.

Dennys, John. *The Secrets of Angling*. London, 1613. STC (2nd ed.) 6611.

Dukas, Reuven, ed. *Cognitive Ecology: The Evolutionary Ecology of Information Processing and Decision Making*. Chicago: University of Chicago Press, 1998.

Fudge, Erica. "The Animal Face of Early Modern England." *Theory, Culture & Society* 30 (2013): 177–198.

Gessner, Konrad. *Historiae animalium, Liber IV, Qui est de Piscium & Aquatilium Animatium natura*. 1558. Frankfurt, 1604.

Gibbs, Raymond W., Jr. "Taking Metaphor Out of Our Heads and Putting It Into the Cultural World." In *Metaphor in Cognitive Linguistics: Selected Papers from the Fifth International Cognitive Linguistics Conference, Amsterdam, 1997*, edited by Raymond W. Gibbs, Jr. and Gerard J. Steen, 145–166. Amsterdam and Philadelphia: John Benjamins, 1999.

Graham, Elspeth. "Ways of Being, Ways of Knowing: Fish, Fishing and Forms of Identity in Seventeenth-Century English Culture." In *Animals and Early Modern Identity*, edited by Pia F. Cuneo, 351–73. Burlington, VT: Ashgate, 2014.

Greenhalgh, Malcolm. *The Pocket Guide to Freshwater Fish of Britain and Europe*. London: Mitchell Beazley, 2001.

Grubb, Thomas C. *The Mind of the Trout: A Cognitive Ecology for Biologists and Anglers*. Madison: University of Wisconsin Press, 2003.

Johnson, Laurie, John Sutton, and Evelyn Tribble, eds. *Embodied Cognition and Shakespeare's Theatre*. New York: Routledge, 2014.

Kilgour, Maggie. "Writing on Water." *English Literary Renaissance* 29 (1999): 282–305.

Kusukawa, Sachiko. "The Sources of Gessner's Pictures for the *Historia animalium*." *Annals of Science* 67, no. 3 (July 2010): 303–28.

Lakoff, George, and Mark Johnson. *Metaphors We Live By*. Chicago: University of Chicago Press, 1980.

Maitland, Peter S. *Biology of Fresh Waters*. New York: Wiley, 1978.

——— and Neville C. Morgan. *Conservation Management of Freshwater Habitats: Lakes, Rivers and Wetlands*. London: Chapman & Hall, 1997.

Mentz, Steve. *At the Bottom of Shakespeare's Ocean*. Shakespeare Now! London: Continuum, 2009.

——— "Brown." In *Prismatic Ecology*, edited by Jeffrey Jerome Cohen, 193–212. Minneapolis: University of Minnesota Press, 2013.

Rondelet, Guillaume. *Libri de piscibus marinis, in quibus veræ piscium effigies expressæ sunt*. Vol. 2, *Uniuersae aquatilium historiæ pars altera*. Lyon, 1555.

Shapiro, Barbara J. *A Culture of Fact: England, 1550–1720*. Ithaca, NY: Cornell University Press, 2000.

Swann, Marjorie. *Curiosities and Texts: The Culture of Collecting in Early Modern England*. Philadelphia: University of Pennsylvania Press, 2001.

*This present boke shewyth the manere of of hawkynge [and] huntynge and also of diuysynge of cote armours* … [The Book of Saint Albans]. Westminster, 1496. STC (2nd ed.) 3309.

Topsell, Edward. *The Historie of Foure-footed Beastes*. London, 1607. STC (2nd ed.) 24123.

———. *The Historie of Serpents*. London, 1608. STC (2nd ed.) 24124.

*Treatyse of fysshynge wyth an Angle*. In *This present boke shewyth the manere of of hawkynge [and] huntynge and also of diuysynge of cote armours* … [The Book of Saint Albans]. Westminster, 1496. STC (2nd ed.) 3309.

Tribble, Evelyn. "Affective Contagion on the Early Modern Stage." In *Affect Theory and Early Modern Texts*, edited by Amanda Bailey and Mario DiGangi, 195–212. New York: Palgrave Macmillan (Springer Nature), 2017. DOI: 10.1057/978-1-137-56126-8.

———. "Distributing Cognition in the Globe." *Shakespeare Quarterly* 56, no. 2 (2005): 135–55.

———. *Early Modern Actors and Shakespeare's Theatre: Thinking with the Body*. London: Bloomsbury Arden Shakespeare, 2017.

Tuana, Nancy. "Viscous Porosity: Witnessing Katrina." In *Material Feminisms*, edited by Stacy Alaimo and Susan Hekman, 188–213. Bloomington: Indiana University Press, 2008.

Walton, Izaak. *The Compleat Angler or, the Contemplative Man's Recreation*. 1st ed. London, 1653. Wing (2nd ed.) W661.

Whitney, Isabella. "The Admonition by the Author to all Young Gentlewomen: And to all other Maids being in Love." In *The copy of a letter, lately written in meeter, by a yonge gentilwoman: to her vnconstant louer* (1597, STC [2nd ed.] 25439). May be found in *Representative Poetry Online*, edited by Ian Lancashire (University of Toronto Libraries, 2003), https://tspace.library.utoronto.ca/html/1807/4350/poem2952.html

———. *The Copy of a letter, lately written in meeter, by a yonge Gentilwoman: to her unconstant Lover. With an Admonition to al yong Gentilwomen, and to al other Mayds in general to beware of mennes flattery*. London, 1597. STC (2nd ed.) 25439.

# 1 Home Ecologies in the *Treatyse of fysshynge wyth an Angle*

The earliest known English text on sportfishing is now missing, but there are two versions of this medieval treatise available for our consideration: an incomplete manuscript from no later than 1450, and a printed copy that found its way into a volume of instructional texts some decades later.[1] Despite a long tradition of attributing the instructions to a person called Juliana Berners (who may or may not have existed), both texts are anonymous.[2] The manuscript (Beinecke MS 171) provides evidence of a distinct fifteenth-century discourse on angling in its own right, separate from works about other sports. The later version does just the opposite, situating angling—at its first emergence into English print—deep within a compilation that treats various forms of recreation and other matters of interest to gentlemen. This chapter considers the 1496 text, entitled the *Treatyse of fysshynge wyth an Angle*, taking it as the more complete rendering of the lost original and as the starting point for a vernacular piscatory literature in England. The *Treatyse* is easily the most important source for much early modern angling instruction, despite the fact that its first printer proclaimed his intent to keep the text hidden from popular view. The documented history of English sportfishing thus begins with explicit efforts to enforce social distinctions, even as the *Treatyse* itself forces readers to think diagonally across environments and species.

In 1486, an anonymous printer at St. Albans assembled a book of instructions aimed at gentlemen; this text, now commonly known as the *Book of Saint Albans*, advertises its anthological nature and lists its contents as "the bokys of haukyng and huntyng with other plesuris dyverse as in the boke apperis and also of cootarmuris a nobull werke."[3] Ten years later, the text was printed at Westminster by Wynkyn de Worde, with a somewhat refined list of contents:

> This present boke shewyth the manere of hawkynge [and] huntynge and also of diuysynge [devising] of cote armours. It shewyth also a good matere belongynge to horses: wyth other comendable treatyses. And ferdermore of the blasynge of armys: as here after it maye appere.[4]

There is no instruction on angling included in the volume from 1486; the fishing treatise tucked into the later version was apparently de Worde's addition, and it may be that he decided rather late in the compiling process to include it. Angling is notably absent from his title, and it isn't mentioned at all until its sudden appearance partway through the treatise on heraldry. The piscatory text is awkwardly inserted, interrupting an otherwise coherent set of information about "cotarmures" and requiring an explanation for its inclusion. Thus, from the get-go, the rhetorical pose of the *Treatyse of fysshynge wyth an Angle* is defensive. Marking the transition from the heraldic discourse (by which "gentylmen shall be knowen from ungentylmen") to the manual on angling, the compiler insists that fishing instruction belongs in "the present volume" because of its intended audience and the particular social force that it can exert upon that audience. Glancing back to remind readers that the book "shewyth afore the manere of hawkynge and huntynge wyth other dyuers maters right necessary to be knowen of noble men," the compiler argues for the inclusion of angling as "one of the dysportes that gentylmen use."

The final reason given for the inclusion of the *Treatyse* hinges on the contrast between angling and other methods of fishing: "And also it is not soo labororyous ne soo dishonest to fysshe in this wyse as it is with nettes & other engynes whyche crafty men done use for theyr dayly encrease of goodes." By framing the sport as both easier and more honest than its nearest recreational relatives, the author begins to portray a practitioner who, despite an obvious interest in efficiency, is never greedy. The word *crafty* will appear repeatedly in the body of the *Treatyse*, always describing the way that piscatorial ingenuity can overcome difficulties in the field. This is the only time it refers to a level of skill that is explicitly correlated with steady material gain ("dayly encrease") and therefore suggests a greater distance from both the nobility of the intended reader and the strictly non-commercial, leisurely pursuit of such a fisher. The pose struck in the transitional note is itself an *engine* designed to catch and hold the attention of an audience that probably isn't expecting a discourse on fishing, so the craftiness of the author is on plain display.

At the end of the *Treatyse*, as if to excuse the lack of advertising in the title, de Worde explains that the knowledge shared in the foregoing pages is not meant for all eyes:

> And for by cause that this present treatyse sholde not come to the hondys of eche ydle persone whyche wolde desire it yf it were enpryntyd alone by itself and put in a lytyll plaunflet therefore I have compylyd it in a greter volume of diverse bokys concernynge to gentyll and noble men, to the entent that the forsayd ydle persones whyche sholde have but lytyll mesure in the sayd dysporte of fysshyng sholde not by this meane utterly dystroye it.
>
> (i4v)

If this is the voice of de Worde, the coda to the *Treatyse* speaks either to his conception of the relationship between textual objects and the maintenance of a social order *or* to his ability to disguise a lack of foresight as a deliberate choice. Either way, the printer introduces the idea that certain contents of the book could be dangerous if they were to fall into the wrong hands, and that he has prevented damage to the "dysporte" itself by refraining from printing a discrete, conspicuous, portable, and presumably more affordable "lytyll plaunflet" on the subject.

In fact, de Worde would go on to print just such a text several decades later—a handsome quarto version of the *Treatyse* was on offer at his Fleet Street shop by about 1532, with the mention of "ydle persones" tellingly omitted.[5] Perhaps his fears about the perils of widespread piscatory knowledge had abated by the thirties, or perhaps they were simply outweighed by a desire to cash in on the increasing popularity of the sport. Certainly the quarto was produced with confidence by a seasoned printer who knew his market. N. F. Blake interprets the printer's relocation at the turn of the century as a decisive change in his enterprise:

> The move reflects the importance of London as a publishing and mercantile centre, and a recognition that Wynkyn had embarked on a new publishing policy. He turned away from the courtly material favoured by [his master] Caxton, which had led him to settle at Westminster, to religious, popular, and educational books, which were better distributed from London.[6]

When de Worde decided to reprint the *Treatyse* as a freestanding text, he had been plying a successful book trade in London for at least two (probably three) decades; the "lytyll plaunflet," with its modernized spelling and increased ornamentation, seems to need no justification of the kind that surrounds its anthologized source in the 1496 *Book of Saint Albans*.

The rhetorical framing devices of de Worde's earlier version—both the transitional note that ushers in the *Treatyse* and the closing comment on the need to keep piscatory knowledge where it belongs—make it clear that angling is a form of social behavior, whatever else it may be. A person who is interested in taking up the sport should, if he follows the implicit advice of the *Treatyse*, ensure that he is sufficiently noble and not too idle to partake in this recreation. Once he has become an angler, he must protect the *game*—in the double sense of the abundant creaturely life that allows the sport to exist, on the one hand, and the challenge and pleasure of attempting to catch that life, on the other: "Also ye shall besye yourselfe to nouryssh the game in all that ye maye: and to dystroye all such thynges as ben deuourers of it" (i4v).[7] In this ambiguous deployment of *game* near the end of the manual, the author calls attention once again to the tangle of contests, between and within species, that constitute the sport.

Through its detailed instructions, the *Treatyse* shows that angling begins at home and that it is, at the start, a readerly undertaking. The author repeatedly advertises the effectiveness of his instructions by promising that the hopeful angler will be successful if he follows them. In this formula, the text itself becomes the teacher: a lost hook is quickly replaced because the good student "maye have store plentee of his owne makynge, as this symple treatyse shall teche hym" (g4v); if one species of fish gets away, the angler will be able to catch another kind "yf he dooth as this treatyse techyth" (g4v). Learning is continually emphasized in the author's preferred categorization of angling as a *craft*—the preamble claims that this text will lay out "the full crafte of anglynge" (h1r), and the instructions that follow frequently echo this term, reminding the reader that his success depends upon the mastery of very specific skills.

Multiple references to *craft* in the *Treatyse*, including the oblique comparison of angling to the skilled labor of tailors and shoemakers, conjure a landscape of learning where information is passed down in text from one person to another, and where two people standing side by side on the riverbank may have drastically different bodies of knowledge. "Yf ye woll be crafty in anglynge," the instructor says, "ye must fyrste lerne to make your harnays" (h1r). The difference between competence and excellence, in this first instance, comes from fabricating one's own gear—the handmade rod is not only crafted but crafty, because its construction is so light and compact: "And thus shall ye make you a rodde soo prevy that ye maye walke therewith: and there shall noo man wyte where abowte ye goo" (h1v). The sport is competitive and resources are limited, so the rod is stealthy. By the river, the angler will craftily deceive fish, but he is first told how to fool his fellow human beings. Both the rod and the instructions that lead to its manufacture are forms of technology that will bring the angler closer to his goal by becoming part of him: the stipulations of the text are internalized and the gear extends the reach of the human body. The expertise and the rod, both formerly personified as external entities, are eventually so intimately wound up with the identity of the angler that they cannot be seen. To be crafty in angling is to be imperceptibly different from—and better than—other people. The attentive reader of the *Treatyse* can refashion himself through a thoughtful recreational practice that will set him apart.

## The Angler's Environments

In *Shakespeare and Ecology*, Randall Martin explains that the origins of localism reach back at least as far as the early modern period: "The cultivation of relatively balanced and genuinely sustainable local relationships ... remains an important conservationist worldview. In early modern

England it was the leading life experience out of which responses to new environmental dangers were conceived."[8] In Martin's reading, Shakespeare's Forest of Arden is an assemblage of "natural and manmade habitats." "Together," he shows us, "these heterogeneous elements form a mosaic: a cluster of patch ecosystems and water and land corridors in a regional area."[9] This description could just as easily be applied to the connected environments of the *Treatyse*, which encourages its readers to wander from farm to forest. The imperative to protect wild habitats through responsible interactions with other species—domesticated and otherwise—is one of the *Treatyse*'s teachings. In a way that "anticipate[d] the bioregionalism of twentieth-century conservation biologist Aldo Leopold and his followers," Martin says, "early modern biodynamic cultivation recognized the need to preserve the physical health of local environments and reaffirmed ancient principles of shared human and non-human dwelling in the world."[10] Although the *Treatyse* does not deal directly with matters of farming, its brand of localism is concerned with the preservation of existing ecosystems, including those which, having been shaped by agricultural and industrial practice, might now be called "novel ecosystems."[11] Like the early modern methods of cultivation in Martin's study, the angler's recreation in the *Treatyse* is a realm of relative responsibility that implicates not only the immediate surroundings of the practitioner but also a "cluster" of places that are sites of belonging for other creatures.

With literature as our portal into early modern angling, it makes little sense to insist on a bright dividing line between nature and culture, or between pristine and manipulated environments. Grappling with the seemingly inevitable "nature-culture distinction," Lawrence Buell employs a flexible definition of "the environment":

> By 'environment(al),' I refer both to 'natural' and 'human-built' dimensions of the palpable world. Though I shall also insist on the distinction, one must also blur it by recourse to the more comprehensive term. Human transformations of physical nature have made the two realms increasingly indistinguishable.[12]

Although it can be difficult to determine where one area ends and another begins, biologists who study "agroecology" see the value in marking a distinction between a "natural ecosystem" and an "agroecosystem." The latter is "a site or integrated region of agricultural production—a farm, for example—understood as an ecosystem."[13] Also "human manipulation and alteration" of existing ecosystems result in substantial differences in energy flow and nutrient cycling, among other factors, but agroecosystems nevertheless *count* as ecosystems and are best understood according to the available models for studying such systems.[14]

The *Treatyse* presents a series of neighboring, imbricate environments: ecosystems that are connected and interpenetrating, bearing to various degrees the signs of human wishes. In addition to the identifiable places that characterize these ecosystems (kitchen, thicket, sheepcote, storefront, gravelly river-bottom) are the many kinds of relationships, within and between species, that give the locations distinct meanings for the reader. Interactions shape ecosystems: smelling the flowers, finding a new fishing hole and keeping it a secret, buying a tailor's needle, clipping a hair from a horse's tail, and quietly reading one's angling manual are all behaviors that have the potential to mold the environments in which they take place. Buell argues for the viability of "a myth of mutual constructionism: of physical environment (both natural and human-built) shaping in some measure the cultures that in some measure continually refashion it."[15] This is the model that best suits a careful reading of the *Treatyse*—or of early angling literature more generally—because it allows us to think of the text itself as a product of animal life, sprung from a particular environment and influencing that environment in turn. Buell would have us think of our sense of place "as containing within it many different patches besides just home, including what comes to us via the world of images as well as through live transactions, plus the changes in us relative to place and the awareness of landscape as timescape."[16] Reading the *Treatyse* and following its advice, the hopeful angler engages with "the world of images" and participates in "live transactions," starting not at the riverbank but long before that, in the ecology of his own home.

The angler's "harnays" is an assemblage of recently living matter, carefully collected from various quarters of the environment surrounding the hopeful fisher's home. These organic materials become the property of the angler through bodily contact and the mental figuring that makes them part of his project. To choose a suitable bough for the purpose of making a rod, one must scout out hazel, willow, and aspen trees between Michaelmas and Candlemas (late September to early February), and find "a fayr staffe of a fadom and an halfe longe" (h1r). Although this length of living wood is already called a "staffe" in anticipation of its future use, the bough will undergo at least a month of treatments in order to become a fishing rod. In the schedule of processes for making this essential tool, the author of the *Treatyse* consistently uses the personal pronoun "hym" to refer to the changing branch:

> Bethe hym in an hote ovyn ... sette hym evyn ... lete hym cole and drye a moneth ... frette hym faste ... bynde hym to a fourme ... brenne hym in the nether ende ... lete hym lye styll and kele two dayes ... Unfrette hym thenne and lete hym drye in an hous roof in the smoke tyll he be thrugh drye.
>
> (h1r-v)

While "he" and "hym" are more versatile in Middle English, not necessarily referring either to people or to grammatically masculine objects,[17] the pronouns do seem to personify the transforming wood in this set of instructions. Elsewhere in the text, "it" is freely used in reference to objects—for example, a mixture of dye (h2r), a hook (h3v), a cork (h4r), and the treatise itself (i4v). Even the rod in question, once it has become a rod, is an "it": "It woll be lyghte and full nymbyll to fysshe with at your luste" (h1v). But during the manipulations that lead to this desired result, the bough is strikingly human as it is bathed, bound, burnt, and left to lie still. The rapidly alternating images of care and torture are all the more arresting for their use of a pronoun that suspends our alienation from the foreign botanical limb and encourages comparisons between human and arboreal tissues. The *it*ness of the cut branch is withheld until we see it fully realized as a portable, practical tool: "your rodde" (h1v). An illustration of the rod appears in the printed *Treatyse*, marking the end of this phase in the instructional text and the beginning of the branch's new, nonliving existence.

In the following section, the angler's line undergoes a similar transformation, beginning as white horse hair and ultimately blending in with the color of the river. Along the way, it ceases to belong to the horse. It takes on a selfhood of its own, then becomes the angler's:

> Fyrste ye must take of a whyte horse taylle the longest heere and fayrest that ye can fynde … Departe it in to vj [six] partes and every parte ye shall colour by hymselfe in dyvers colours. As yelowe: grene: browne: tawney: russet, and duske colours. And for to make a good grene colour on your heer ye shall doo thus.
>
> (h1v-h2r)

The segmented hair is a series of selves, each of which will be manipulated in a particular way so as to achieve a certain color. In the dyeing instructions, the fiber from the horse's tail is consistently called "your heer," a formulation that would be confusing if horse hair and human hair were understood to belong in the same category. Instead, it's obvious that the equine hair has a certain flexibility; it can move from one owner to another without losing its essential properties. Choosing to repeat the word "heer" rather than switching to the less affiliated "line," the author again preserves a sense of the object's living origin.[18]

Like so much of the tackle described in the *Treatyse*, both the rod and the line begin as part of living organisms and are gradually divorced from their points of origin through manipulations that include changes in terminology. The altered branch is not a rod until it is so called, and the horse's hair must become "your heer" before it can be treated as a length of fishing line. Language is one of the things the angler applies to these objects in order to make them part of his "harnays." The shifts in diction, along with gaps in the instructional fabric, point up the particularity of the relationships that already inhere between the fisher and other organisms. The tail of the horse comes readily to hand, and there's no map showing the way to the nearest copse of hazel trees. The author takes for granted the availability and familiarity of many kinds of living resources within the reader's known environment, and these assumptions reach an impressive pitch in the patterns for artificial flies given late in the *Treatyse*. The instructions proceed month by month, artfully combining the materials that will best approximate seasonal bugs; for example, "In the begynnynge of May a good flye[:] the body of roddyd wull and lappid abowte with blacke sylke; the winges of the drake and of the redde capons hakyll" (i3v). To make a fly that really convinces in early May, one needs not only to have paid attention to the insect life commonly represented at the river at that time of year; the angler also requires products from a sheep, a worm, a duck, and a rooster to create a reasonable facsimile of such a creature.

So the world imagined around the reader of the *Treatyse* includes not only a wilderness in which one can forage and fish; there's also a domestic space that is shared by various animal species and the human beings who built it. Some of these—like the capon and sheep whose feathers and fleece become part of the artificial fly—are kept deliberately adjacent to (if not within) the human dwelling. Even if the reader keeps no sheep, raises no chickens, and has no white horse, he is assumed to live within easy distance of people who do. Furthermore, the text reveals that the border between wild space and the *oikos* of the human home is shifty and permeable; when the angler takes his carefully crafted fly to the riverbank, he mingles materials from his own habitat with that of the fish. Similarly, the rod, once a living limb on another body, becomes an extension of the fisher's arm, allowing him to enter the water by proxy.

A full picture of the *Treatyse*'s ecology cannot stop, however, at the extraordinary variety of nonhuman life used by the angler. Relationships among people are also essential components of the angler's world, since they enable him to gather the necessary gear. Aside from the animal products that one's neighbors might have at hand, there are human-made items that can be put to new use. In order to form hooks ("the moost subtyll and hardyste crafte in makynge of your harnays") the angler needs a substantial store of tools, including a hammer, a clamp, a file, and an anvil. The hooks themselves begin as needles of various sizes from a range of trades. For the smallest fish, one needs "the smallest quarell [square] nedlys" (h3r); perhaps these were similar to those used in surgery.[19] Larger hooks are made from "greter nedles: as broderers nedlis: or taylers: or shomakers nedlis" (h3r). The sturdiest begin as "shomakers nalles" (h3v). Rather than making a set of hooks from scratch, the angler can draw from the supply of specialized tools that have been fabricated for use elsewhere, and alter them so that they will be suited to his recreational purposes. The text here echoes the temporary personification of the rod-making instructions—"Thenne alaye [allay or temper] hym again:

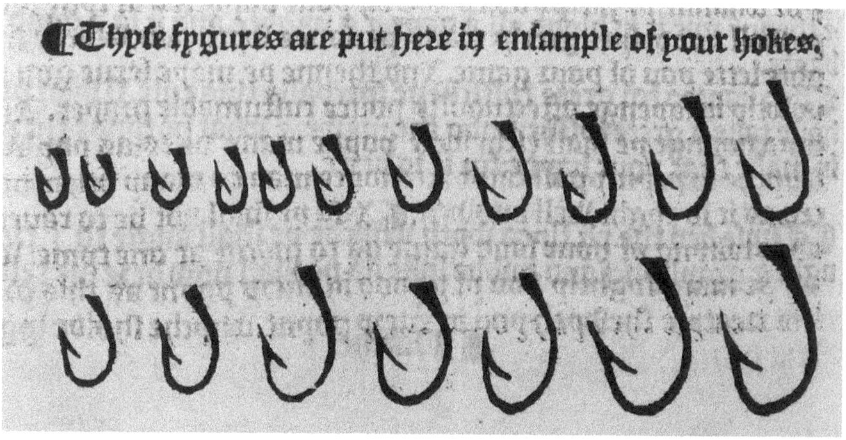


*Figure 1.1* Fish hooks from the *Treatyse of fysshynge wyth an Angle*, 1496. Beinecke Rare Book and Manuscript Library, Yale University.

for elles he woll breke in the bendyng" (h3r)—enforcing the idea that the needle is leaving one state of being and entering another. Although the original objects are not depicted, the author directs the reader to an image that appears several pages later, showing the appropriate curvature of hooks in many sizes: "Thenne bende hym lyke to the bende fyguryd herafter in example" (h3r).

The bending of needles that are meant for one purpose into hooks that have an entirely different orientation is a tidy representation of how the *Treatyse* situates the sport of angling within a social world. As he heats, bends, and sharpens the tools of someone else's trade, the fisher is dependent upon an existing environment of human commerce—an economy that is, in turn, part of an ecology. The opportunistic use of shoemakers' needles is, after all, not so different from the harvesting of a hazel bough or the plucking of duck feathers. Urban life, and the crafts that flourish therein, must be counted among the biotic phenomena that surround the angler. With materials gathered from woods, stables, and shops, the reader will travel to yet another neighboring environment and perform yet another kind of animal activity.

The angler's place in a mosaic of ecosystems (to use Randall Martin's metaphor) is perhaps most plainly illustrated in the woodcut on the frontispiece of the *Treatyse*, which shows the rocks, plants, and water of the riverside in the foreground—not to mention the fish being lifted out of the stream—and an urban settlement in the distance. The fisher's body is oriented wholly toward the river; with his back to the town, he faces the fish. One hand reaches toward the catch at the end of his line, while the other holds his rod parallel to the diagonal line of the river's edge. These diagonals out front contrast with the perpendicular plane of the buildings behind, so the geometry of the composition reinforces the distinction between wild and peopled space. The angler has positioned himself in such a way that we can apprehend a resemblance between the human being and the fish he catches. The two bodies are similarly curved, and each is divided into sections, with the more decorative side on the left. The fish's lateral line is like the placket down the angler's middle, and where the man wears buttons, the fish wears spots. To an extent, the aquatic creature looks like a miniature version of the terrestrial. At the same time, however, the human seems to match the town rather than the stream: the buttons of his doublet and the cap on his

*Figure 1.2* The frontispiece from the *Treatyse of fysshynge wyth an Angle*, 1496. Beinecke Rare Book and Manuscript Library, Yale University.

head make him visually similar to the windowed, peaked buildings in the distance, while the fish is neatly camouflaged by its home waters. Multiple concurrent relations are perceptible within the single, small frame of the woodcut.

The proper goal of the angler, according to the logic of the *Treatyse*, is not to catch a lot of fish (or even one), but to be immersed in a nearby environment that is partly foreign. The pleasure of the sport is derived in part from its challenges—as the supreme leisure activity, sportfishing must entail some element of difficulty without making its practitioner ill, as other endeavors are said to do. After constructing at home the elaborate equipment that angling requires, the fisher has to study his surroundings and place his body strategically in relation to his prey. To this end, the *Treatyse* provides a catalog of different "place[s] of the water" and indicates which of these are best for angling. By reading the landscape and choosing a spot, the sportfisher sets the level of his own challenge.

The list of potential piscine habitats includes places that combine human-built and natural phenomena in a way that is typical of the *Treatyse*: "It is good for to angle in depe styffe stremys and also in fallys of waters and weares: and in floode gatys and mylle pyttes" (h5r); fish may "have restynge behynde stangnys and archys of brydgys and other suche manere places" (h5v). Naming hydrological structures in tandem with rivers and pools, this accounting of potential fishing spots tacitly acknowledges the fact that human interventions can create, destroy, or alter the opportunities of other creatures within an existing environment, and that people can, in turn, learn to adjust their behavior according to changes in the shape and life of a place. A fish does not have to recognize a millpond as a store of industrial power in order to benefit from its existence, any more than human beings need to know why it is comfortable to rest in the waters under a bridge. "Suche manere places" are inevitably understood differently by all who visit them, and may not even register as distinct locations. As Lawrence Buell puts it, "Historically, what different groups of humans (and nonhumans) reckon to be a meaningful 'place' may diverge and crisscross in innumerable ways. What feels like a place to me may not feel so to you, and likewise with whole cultures."[20] The good angler strives to find piscine meaning in nearby places, setting aside the human signification of weir, floodgate, mill, and bridge, and observing the continuity—so perfectly embodied by the swimming fish—from one potential dwelling to the next.

Pools of standing water are easier to fish because "there is not grete choyse of ony places where it is ony thynge deepe in a pole" (h5r)—this dearth of options refers not only to the angler's vetting of positions, but to the fish's search for a safe and sustaining place to dwell. "For it is but a prison to fysshe," the instructor says, "and they lyve for the more parte in hunger lyke prisoners: and therefore it is the lesse maystry to take theym" (h5r). Unable either to hide or to resist the temptation of bait when it is offered, such vulnerable fish are not especially satisfying to catch. The greater craft is required at the river, where one must choose from many different deep places. One can look for spots where "it is depe and clere by the grounde: as gravel or claye without mudde or wedys," but these will not necessarily prove the most bountiful. In an eddy, in a cavity along the bank, in a tangle of tree roots, or in a group of weeds swishing in the water, "the fysshe maye covere and hyde themselves at certain tymes whan they lyste" (h5r). Unlike their relatives in nearby pools, these running-water fish may come and go from a particular place as they please. The trained angler searches not for the clearest view but the most likely abode, keeping in mind the caprice of his prey. The "certain tymes" our author mentions are not certain at all—they are unpredictable, determined only by the fish's whim. So it is not only the verdant growth in and near the river, or the complex, swirling course of the water that makes the angler's recreation challenging and fun; it's also, centrally, the alien inclinations of the fish—the way they "lyste"—that allows the human on the bank to experience befuddlement or vindication, depending on the day.

The tantalizingly hidden fish are analogous, in their preference to remain out of view, to the angler who can make his way to the riverbank without showing any evidence of his intention to fish. Armed with subtle gear and invisible know-how, the reader of the *Treatyse* can move unseen from one habitat to another. If it seems advantageous for the angler to go unnoticed by members of his own species, it is absolutely crucial that he is not noticed by fish. In fact, according to the *Treatyse*, this is the number-one thing:

> And for the fyrste and pryncypall point in anglynge: kepe thee ever fro the water fro the sighte of the fysshe: other feare on the londe: or ellys behynde a busshe that the fysshe se you not. For yf they doo they wol not byte.
>
> (h4v)

The hopeful fisher should not even cast a shadow on the water, "for it is that thynge that woll soone fraye the fysshe" (h4v). The *Treatyse*, at least in its first printed iteration, privileges sneaking about and hiding as the characteristic behaviors of angling on all sides: the fish lurk under weeds and structures, the fisher escapes the view of both his prey and his competitors, and the compiler of the text keeps these instructions out of the sight of "ydle persones" by stashing them in a large collection of works on other subjects.

## Spirited Ecology

Throughout, the *Treatyse* describes a world of competition. Angling is, of course, figured as a contest between the individual fish and the person trying to catch it, who vies, in turn, against any other anglers who may be fishing the same stream. But the sport itself is also represented as a competitor: in the author's opening gambit, angling is pitted against hunting, hawking, and fowling, and emerges victorious—it is, according to the *Treatyse*, "the very meane and cause that enducith a man in to a mery spyryte" (h1r), while the other sports leave their practitioners cold, wet, dirty, and disappointed. Beyond this comparison to sports that entail the training and pursuit of aerial and terrestrial animals, angling is also differentiated from other methods of fishing. These, too, result in dangerously sodden clothes, while the technology of the rod presumably keeps the angler warm and dry. "All other manere of fysshyng is also laborious and grievous," the author declaims, "often makynge folkes ful wete and colde, whyche many tymes hath be seen cause of grete Infirmytees. But the angler maye have no colde nor no disease nor angre but yf he be causer hymself" (g4v). The sport is a low-risk exercise in self-sufficiency; the angler might lose a hook or a line, but can easily replace these from his plentiful store of homemade gear. His sport wins because it is a game that is impossible to lose.[21]

Even the very worst that can happen—a total failure to catch fish—is no big deal, because there is much more to be gained from an excursion to the riverside. The spiritual benefits of the sport accrue whether or not any fish is brought to land:

> And yet atte the leest [the angler] hath his holsom walke and mery at his ease. A swete ayre of the swete savoure of the meede floures: that makyth hym hungry. He hereth the melodious armony of fowles. He seeth the yonge swannes: heerons: duckes: cotes and many other foules wyth theyr brodes. Whyche me semyth better than alle the noyse of honndys [hounds]: the blastes of hornys and the scrye of foulis that hunters: fawkeners and foulers can make.
>
> (g4v-h1r)

The angler hears no noise, only music. Ironically, the plainly antagonistic pursuit of fishing wins the title of most wholesome sport, not by guaranteeing a substantial quarry or perfect ease of execution, but by placing its practitioner in a state of ecological contemplation where his body becomes a porous recipient of news from other species: the aromas sent up by wildflowers, the songs of birds, the spectacle of cygnets, and ducklings with their watchful parents. To rise early and visit the river is to take care of one's soul by opening up to an array of sense impressions.

In the *Treatyse*, angling is an opportunity to witness and feel what Stacy Alaimo calls *trans-corporeality*: a complex of interactions "in which the human is always intermeshed with the more-than-human world."[22] The "ayre" and "armony" of the angler's environment do not remain without, but enter his body and take effect. In Alaimo's terms, "The 'environment' is not located somewhere out there, but is always the very substance of ourselves;" ... "bodies and places are continuous."[23] Instead of detailing the joys of dining on the day's catch, the *Treatyse* describes how angling might cultivate in its practitioner a sustaining *hunger*. When the bodies of wildflowers express themselves through smell, the angler's body absorbs this information and becomes hungry. That experience of yearning is presented as a good in itself, a transgression of bodily boundaries that allows the angler to feel fully integrated—if only temporarily—into the patch ecosystem that is home to the fragrant blooms. Leaving his house and his town, he finds himself among botanical and avian neighbor species who quietly induct him into a devotional practice.

The option of isolating oneself from other people is an important dimension of sportfishing. Without stipulating that an angler must fish alone, the author indicates that it's best to go solo. Toward the end of the *Treatyse*, the solitary nature of the sport is extolled as part of its spiritual wholesomeness. In an "order" to readers, declaring how the foregoing instructions are to be used, the author lays out several moral principles: Do not fish in any water that is owned by a poor man; do not break anyone's trap or steal the fish therein; do not destroy another man's hedges as you make your way to the river. The final item in this list is an injunction against angling solely for monetary gain:

> Also ye shall not use this forsayd crafty dysporte for no covetysenes to thencreasynge and sparynge of your money oonly, but pryncypally for your solace and to cause the helthe of your body and specyally of your soule. For whanne ye purpoos to goo on your disportes in fysshyng ye woll not desyre gretly many persones wyth you. Which myghte lette you of your game. And thenne ye maye serve god devouwtly in sayenge affectuously youre custumable prayer.
>
> (i4r)

The competition for limited resources—the same factor that led to the fabrication of the lightweight, top-secret rod—compels the angler to find a private spot where no other crafty sportsmen will interfere with his catch. Solitude, in turn, allows the angler to issue the kind of praise that comes most naturally to him: his "custumable prayer," the one he usually utters. In other words, he is able to be himself, before God. Instead of hampering the spiritual health of its practitioners, angling makes a virtue of competition by placing the contenders at a quiet, contemplative distance from one another. To see clearly the ecosystems described in the *Treatyse* is to envision not only the tangled relationships between an angler, his fellows, and the fish they attempt to catch, but also the celebrated intimacy between a person and their god. The ecology of the *Treatyse* is not secular but spirited; its ecosystems include a watchful divine power.

The injunctions laid out at the end of the *Treatyse* prescribe a pattern of neighborly conduct that extends to fellow human beings and beyond.

In this final suite of rules comes an explicit reminder of the manual's intended audience: "noble men" are mentioned repeatedly, in contrast with "poore" men, "theuys [thieves]," "brybours [bribers]," and "ydle persones" (i4r-v). After being warned about the evils of stealing fish and damaging property, the *Treatyse*'s gentle reader is reminded once again about the spiritual benefits of angling, and finally given a lesson in sustainability: "Also ye shall not be to ravenous in taking of your sayd game as to moche at one tyme" (i4r). Although the author claims that the advice in the *Treatise* will easily afford the hopeful angler an overabundance of fish, he must restrain himself—for his own good, for the benefit of his fellow anglers, and for the survival of the sport. To do otherwise would be to "dystroye your owne dysportes and other mennys also" (i4v). Instead of overfishing, the responsible angler works to sustain riparian life: "Ye shall besye yourselfe to nouryssh the game in all that ye maye: and to dystroye all suche thynges as ben deuourers of it" (i4v). The rod-toting angler apparently does not number among such "deuourers;" on the contrary, he is in a position to defend piscine life from a range of unspecified dangers. With this urging to protect the river and its creatures comes a potential benediction: "And all those that done after this rule shall have the blessinge of god and saynt Petyr, whyche he theym grante that with his precious blood us boughte" (i4v). Christ's sacrifice emerges here as a reminder and a reward for proper piscatorial conduct: if you follow this commandment by nourishing the fish, you in turn will be nourished by divine agency. It also sounds like an exaggeration of the sportfisher's own sacrifice; while the angler lets many fish stay alive in their habitat, Christ purchases the salvation of all humankind. The *Treatyse* thus finds a way to make the sportfisher appear as a good Christian neighbor to his prey.

## Feeding Fish and Bringing Bugs Home

At its core, the *Treatyse* is a field guide to desires, and yet its primary focus is never the human appetite. There is not a single piece of advice about how to prepare fish for the table, despite the fact that this must be the culminating point of the angler's practice; rather, we find dozens of recipes for bait. And in the inverted telescope of the *Treatyse*'s comprehensive investigation, the story doesn't end with the gaping maws of duped fish. The bait, too, must be fed. Thus, the text represents an ecology of angling that can be largely understood through information about food sources. Stacy Alaimo suggests that "Perhaps the most palpable trans-corporeal substance is food, since eating transforms plants and animals into human flesh."[24] The *Treatyse* gives plenty of time to questions of nutrition, making food a major category of inquiry even as it decenters human ingestion. Instead of serving up recipes for the angler to enjoy, it describes an ecosystem that is made up of diverse metabolisms, and informs the reader about hungers that originate in other bodies. In its systematic delivery of instructions, the *Treatyse* illuminates not only the transit between human and non-human bodies, but also the intermediary roles human beings might play in the relationships between other species. Trans-corporeality is a useful concept for thinking about the capture and consumption of fish; it can be even more compellingly applied to the preparation and delivery of foods *for* fish. While the process of eating a trout can turn the fish into human flesh, *feeding* a trout—or seeming to—surely transforms the serving body in another, perhaps more subtle way.

Like many of the fishing manuals that were composed after it, the *Treatyse* organizes its advice into short discourses on individual species. The fish are often described with reference to their flavor, texture, and nutritional value: the grayling is "a delycyous fysshe to mannys mouthe" (h6v); the barbel is "swete," but "a quasy meete and a peryllous for mannys body" (h6v); the ruffe is "ryght an holsom fysshe" (i2r). But these observations about human consumption are given very little space relative to the extensive information about the feeding habits and gustatory preferences of fish. Many introductions to species begin with a comment on how the fish tastes to humans, then shift rapidly to a description of how the fish tastes its bait: "The Troughte ... is a right deyntous fyssh and also a ryght fervente-byter" (h6r); "The flounder is an holsom fishe and a free, and a subtyll byter in his manere" (i2r); "The menow whan he shynith in the water thenne is he byttyr And though his body be lytyll yet he is a ravenous biter and an egre" (i2v). The purpose of these short natural histories is to instruct the reader in appropriate offerings for each species, because, as the author points out, "ye can not brynge an hoke in to a fyssh mouth without a bayte" (h6r).[25]

The *Treatyse* thus promises to deliver advice on baits along two axes: "for every manere of fyssh and for every moneth" (h6r). The menu for the roach, for example, is a staggering array of locally sourced treats:

> In Marche the most redy bayte is the red worme. In Apryll the bobbe [grub] under the cowe torde. In May the bayte that breedyth on the oke leyf and the bobbe in the dunghill. In June the bayte that bredith on the osyer and the codworme. In Juyll hous flyes and the bayte that bredith on an oke, and the notworme and marhewes and maggots tyll Myghelmas. And after that the fatte of bakon.
>
> (i2r)

In the fabric of these suggestions is a negotiation between what is seasonally available ("most redy") and what pleases the fish in question. Both delineations require generalization—about the ecological patterns that mark the passage of the seasons on the one hand, and about the typical habits of individuals within a species on the other. The *Treatyse* establishes a dietary calendar or almanac, not unlike the projected list of contents you might find in an advertisement for a CSA (Community Supported Agriculture group) share: tomatoes in August, parsnips in November, kale all the time.

More precisely, the harvesting program of the *Treatyse* resembles several tables in the United Nations' 2013 report on edible insects, which have months listed in one column and available creatures listed in the next. In the table entitled "Availability of edible insects in Thailand, by month," for example, we find that the "Giant water bug, wood-boring beetle, [and] predaceous diving beetle" are abundant in June; in July, one can gather the "Back swimmer, crawling water beetle, damselfly, [and] spider."[26] The common names listed in such catalogs are predominantly relational, like those in the *Treatyse*—they identify species according to observed habitats and behaviors, indicating a familiarity with the morphology and daily activity of each animal. "In Mexico," the report reads:

> Indigenous peoples possess a deep knowledge of the plant and animal species that traditionally make up their diets, including life cycles of insects. Insects have been 'calendarized' by species, meaning that they are believed to operate in harmony with natural phenomena such as plant life cycles, moon cycles, rainy seasons and thunder. It is widely known among indigenous people, for example, that escamoles (larvae of the ants of the *Liometopum* genus) are ready to harvest when the jarilla plant (*Senecio salignus*) is flowering.[27]

The *Treatyse* likewise acknowledges the spatial and temporal aspects of habitat, attending to reproductive cycles and other seasonal behaviors. In the manual's taxonomy of bait, insects have home addresses near the dwelling of the angler. The names of the creatures in question are in many cases also their habitats, which are often the bodies and products of other organisms: for example, "the bobbe under the cowe torde," "a flesshe flye," "the bayte that bredyth betwene the tree and the barke of an oke," and "the water docke leaf worme" (h6v-i1r). These labels locate insect species within the region of patch ecosystems where the reader of the *Treatyse* is expected to live, and they sometimes identify relationships between vertebrate and invertebrate species that would not have come about without human-made changes to existing environments ("the bobbe under the cowe torde" is a good example).

Bugs who rely on the bodies or waste products of livestock can be considered *commensal* with human beings, since they thrive within an anthropic (human-built) environment. Remembering that it can be difficult to distinguish between natural and human-made patches, we can think of coprophagous insects—in the *Treatyse* and beyond—as representatives of sustainable agroecological relationships. Archaeologist Terry O'Connor cites the example of livestock feces that appear on farms as a result of the human manipulation of wild space (and wild species), and in turn can create opportunities for new commensal relationships: "Traditionally maintained farmyards often have accumulations of herbivore dung and decomposing plant matter that become an attractive feeding and breeding pabulum for flies."[28] Both the flies and the animals who visit the farm to prey upon them are part of "the commensal guild"—the highly varied cluster of species "that utilize the modified or constructed environment of human habitations for food and living space."[29] O'Connor emphasizes the importance of food in commensal relationships by reminding us that the word itself literally refers to the sharing of a table;[30] although our food is not always identical, humans and other species often eat together.

Promoting entomophagy for the purpose of achieving global food security, the UN report explains that "Most edible insects are harvested in the wild," and that long-standing practices of foraging can be extended into the realm of farming: "Insect gathering and rearing as minilivestock at the household level or industrial scale can offer important livelihood opportunities for people in both developing and developed countries."[31] The report also describes the sustainable use of insect protein in "feed"—food for other, non-human animals[32]—and draws attention to the benefits to be had from opportunistic species that appear in anthropic environments. Here, too, coprophagous species get special mention: "Dung beetles—of which there are about 4,000 known species—play a significant role in decomposing manure."[33] In general, then, the report shares an orientation with the *Treatyse*, looking toward insects as nifty, edible entities that can be harvested close to home.

The angler has to visit habitats adjacent to his own—within his home ecosystem—and interrupt life cycles that are much less visible than the breeding of cows or sheep (or fish, for that matter) in order to procure the best bait. This is not a hunt, exactly, but it is a pursuit that requires precise attention to environments on a minute scale. Long before he reaches the riverbank with its various hiding places for trout and salmon, the sportfisher must imagine, find, and plunder the homes of insects. Another thing that distinguishes the collection of bait from the capture of fish is that many of the insects are gathered alive and kept that way. These "quycke baytes" are domesticated, in the sense that a recognizable home is established for them within the angler's dwelling: "Ye shall fede and kepe them all in generall: but every manere by hymself wyth suche things in and on whiche they brede" (i3r). Maggots are an exception, as they are treated to an unusually luxurious diet and a warm rest:

> Whan thei ben bredde grete wyth theyr naturell fedynge: ye shall fede theym furthermore wyth shepes talow & wyth a cake made of floure & hony. Thenne woll they be more grete. And whan ye have clensyd theym wyth sonde in a bagge of blanket kepte hote under your gowne or other warm thyng two houres or thre, then ben they beste and redy to angle wyth.
>
> (i3r)

These animals certainly belong in the UN report's category of minilivestock, as they are valuable living possessions that share the home of their breeder. They are commensal with their owners to a striking degree, now being fed from the store of foods that would otherwise be intended for human consumption.

The quick baits of the *Treatyse* can also be placed under the rubric "living property," a new classification that, according to the animal law scholar David Favre, is necessary for animals' emergence into "legal visibility."[34] Unlike personal property (your car, your shoes, your apple tree), people would generally agree of domestic animals that their human owners have "a duty toward [their] well-being."[35] But as Favre acknowledges, insects present a particular problem for the field of animal law. Although the most fundamental distinction between my electric mixer and my dog is that my dog is alive, life itself is not sufficient to demand rights—few would argue for greater legal visibility of the apple tree, never mind the *Phyllachora pomigena* colonies living on the skin of its fruits. It's more relevant that one's dog has "self-interests and a desire for continued life," in addition to the ability "to experience pain in roughly the same way that humans experience pain;" these are the distinctions that Favre signals when he limits his analysis to the potential legal status of vertebrates, explicitly excluding arthropods from the equation: "Insects can be put off until another time." So distant are insects from membership in the new category of living property that it's possible for Favre to use them metaphorically: "As the chrysalis must leave its cocoon to be the butterfly, so animals must leave the category of personal property to move into their next stage within the legal systems of the world."[36] No thought, then, for the legal status of the caterpillar, chrysalis, or butterfly.

## Neighborly Disgust

It is perhaps too much to ask that insects be granted the kind of legal visibility that has only recently been considered for species more closely related to humans, but we can nevertheless recognize that the author of the *Treatyse* demonstrates an admirable comfort in the company of arthropods. Like the UN report on entomophagy, the late medieval fishing manual is geared toward the use of insects as food; it is not concerned with preserving these tiny lives for their own sake, but with garnering them for human ends. It is striking, though, that there should be so little room in the *Treatyse* for disgust, a feeling that is often exemplified by human responses to insects. The UN report explains that disgust is a cultural barrier to the promotion of entomophagy.[37] Martha Nussbaum notes that "insects are … especially likely to be linked with the disgust-properties that signal decay—stickiness, sliminess and other signs of our animal mortality and vulnerability."[38] In an effort to document the "cognitive content" of this feeling, Nussbaum cites the work of Winifred Menninghaus, who "speaks of disgust as a 'crisis

(2nd ed.) 3309. Unless otherwise stated, all quotations from the *Treatyse* are drawn from this volume. I retain the spelling and punctuation of this edition, but shorten long S and expand superscripts.

5 The quarto is held at the Morgan Library in New York City (accession number PML 20894), where the catalogue suggests a date of ca. 1532–1534. Although there are many differences between this freestanding edition and the earlier text, it is clear that de Worde was working from his own version of 1496.

6 N. F. Blake, "Worde, Wynkyn de (d. 1534/5), printer," *DNB* (January 2008), DOI: https://doi.org/10.1093/ref:odnb/29968.

7 *Middle English Dictionary*, s.v. "game, n.," senses 2c and 6, accessed April 6, 2018, http://quod.lib.umich.edu/cgi/m/mec/med-idx?type=id&id=MED18146.

8 Randall Martin, *Shakespeare and Ecology* (Oxford: Oxford University Press, 2015), 33.

9 Martin, *Shakespeare and Ecology*, 18.

10 Martin, *Shakespeare and Ecology*, 19.

11 Nathaniel B. Morse, Paul A. Pellissier, Elisabeth N. Cianciola, Richard L. Brereton, Marleigh M. Sullivan, Nicholas K. Shonka, Tessa B. Wheeler, and William H. McDowell, "Novel ecosystems in the Anthropocene: a revision of the novel ecosystem concept for pragmatic applications," *Ecology and Society* 19, no. 2 (2014): 1.

12 Lawrence Buell, *Writing for an Endangered World: Literature, Culture, and Environment in the U.S. and Beyond* (Cambridge, MA: Harvard University Press, 2001), 6, 3.

13 Stephen R. Gliessman, *Agroecology: The Ecology of Sustainable Food Systems*, 2nd ed. (Boca Raton, FL: CRC Press, 2007), 23.

14 Gliessman, *Agroecology*, 29.

15 Buell, *Writing for an Endangered World*, 6.

16 Buell, *Writing for an Endangered World*, 77.

17 *MED*, s.v. "he, pron.(1)," accessed April 7, 2018, http://quod.lib.umich.edu/cgi/m/mec/med-idx?type=id&id=MED20189; *OED*, s.v. "he, pron., n.1, and adj.," sense A.I.2.b, revised September 2013.

18 Incidentally, the process by which the angler comes into possession of his gear in the *Treatyse* matches up with John Locke's labor theory of property: "The *labour* of [a person's] body, and the *work* of his hands, we may say, are properly his. Whatsoever then he removes out of the state that nature hath provided, and left it in, he hath mixed his *labour* with and joined to it something that is his own, and thereby makes it his property" (*Second Treatise of Government: An Essay Concerning the True Original, Extent and End of Civil Government*, ed. Richard Cox [Wheeling, IL: Harlan Davidson, 1982], 18. This edition is based on a hand-corrected copy of the 1698 printing of Locke's *Two Treatises*.)

19 Juhani Norri, *Dictionary of Medical Vocabulary in English, 1375–1550: Body Parts, Sicknesses, Instruments, and Medicinal Preparations* (New York: Routledge, 2016), 712 (s.v. "quarrel needle").

20 Buell, *Writing for an Endangered World*, 60.

21 A defensive approach is typical of late medieval treatises on hawking and hunting, as John McDonald shows (*The Origins of Angling*, 14–17), so it isn't surprising to find the author of the *Treatyse* arguing for angling as a virtuous alternative to all other sports.

22 Stacy Alaimo, *Bodily Natures: Science, Environment, and the Material Self* (Bloomington: Indiana University Press, 2010), 2.
23 Alaimo, *Bodily Natures*, 4, 11.
24 Alaimo, *Bodily Natures*, 12.
25 The word *bait* comes down to the author of the *Treatyse* partly from a pair of Old Norse words that refer to the human orchestration of nonhuman eating: the neuter noun *beit* referred to pasture, and the feminine noun *beita* meant simply "food," but had the coloring of a morsel set out to lure in unsuspecting prey. The Middle High German word *beize* meant "hunting" (*OED*, s.v. "bait, n.1.," accessed April 7, 2018, www.oed.com/view/Entry/14737?rskey =buNH6X&result=1&isAdvanced=false). As verbs, *bait* and *bite* share their derivation, but by the fourteenth century the words were distinct enough that to *bayte* meant specifically to make one animal bite another (*OED*, s.v. "bait, v.1.," accessed April 7, 2018, www.oed.com/view/Entry/14739?rskey=buNH 6X&result=3&isAdvanced=false).
26 Arnold van Huis et al., *Edible insects: future prospects for food and feed security* (FAO Forestry Paper 171) (Rome: Food and Agriculture Organization of the United Nations, 2013), 18.
27 van Huis et al., *Edible insects*, 18.
28 Terry O'Connor, "Commensal Species," in *The Oxford Handbook of Animal Studies*, ed. Linda Kalof (Oxford: Oxford University Press, 2014), accessed May 5, 2018, DOI: 10.1093/oxfordhb/9780199927142.013.004.
29 O'Connor, "Commensal Species."
30 O'Connor, "Commensal Species."
31 van Huis et al., *Edible insects*, xiv, xv.
32 van Huis et al., *Edible insects*, xiv, 25.
33 van Huis et al., *Edible insects*, 5.
34 David Favre, "Animals as Living Property," in *The Oxford Handbook of Animal Studies*, ed. Linda Kalof (Oxford: Oxford University Press, 2015), accessed May 5, 2018, DOI: 10.1093/oxfordhb/9780199927142.013.14.
35 Favre, "Animals as Living Property."
36 Favre, "Animals as Living Property."
37 van Huis et al., *Edible insects*, 35–36.
38 Martha Nussbaum, *Hiding from Humanity: Disgust, Shame, and the Law* (Princeton, NJ: Princeton University Press, 2006), 86.
39 Nussbaum, *Hiding from Humanity*, 82.
40 Nussbaum, *Hiding from Humanity*, 91.
41 Sigmund Freud, *Civilization and Its Discontents* (1929), quoted in Slavoj Žižek, Eric L. Santner, and Kenneth Reinhard, *The Neighbor: Three Inquiries in Political Theology* (Chicago: University of Chicago Press, 2010), 2.
42 Žižek, Santner, and Reinhard, *The Neighbor*, 7. The hope of these essays is "to make psychoanalysis a key resource in the project of reanimating the ethical urgency and significance of neighbor-love in contemporary society and culture" (2).

## References

Alaimo, Stacy. *Bodily Natures: Science, Environment, and the Material Self*. Bloomington: Indiana University Press, 2010.

Blake, N. F. "Worde, Wynkyn de (d. 1534/5), printer." In *Oxford Dictionary of National Biography*. Oxford University Press, January 2008. DOI: https://doi.org/10.1093/ref:odnb/29968.

Boffey, Julia. "Berners [Bernes, Barnes], Juliana (fl. 1460), supposed author and prioress of Sopwell." In *Oxford Dictionary of National Biography.* Oxford University Press, September 2004. DOI: https://doi.org/10.1093/ref:odnb/2255.

Buell, Lawrence. *Writing for an Endangered World: Literature, Culture, and Environment in the U.S. and Beyond.* Cambridge, MA: Harvard University Press, 2001.

Favre, David. "Animals as Living Property." In *The Oxford Handbook of Animal Studies*, edited by Linda Kalof. Oxford: Oxford University Press, 2015. DOI: 10.1093/oxfordhb/9780199927142.013.14.

Freud, Sigmund. *Civilization and Its Discontents.* 1929. Quoted in *The Neighbor: Three Inquiries in Political Theology*, by Slavoj Žižek, Eric L. Santner, and Kenneth Reinhard, 2. Chicago: University of Chicago Press, 2010.

Gliessman, Stephen R. *Agroecology: The Ecology of Sustainable Food Systems.* 2nd ed. Boca Raton, FL: CRC Press, 2007.

*Here begynneth a treatyse of fysshynge wyth an Angle.* London, ca.1532–34. Morgan Library accession no. 20894.

*Here in thys boke afore ar contenyt the bokys of haukyng and huntyng with other plesuris dyuerse as in the boke apperis and also of cootarmuris a nobull werke* [*The Book of Saint Albans*]. St. Albans, 1486. STC (2nd ed.) 3308.

Locke, John. *Second Treatise of Government: An Essay Concerning the True Original, Extent and End of Civil Government*, edited by Richard Cox. Wheeling, IL: Harlan Davidson, 1982.

McDonald, John. *The Origins of Angling*. New York: Doubleday, 1957.

Martin, Randall. *Shakespeare and Ecology.* Oxford: Oxford University Press, 2015.

Morse, Nathaniel B., Paul A. Pellissier, Elisabeth N. Cianciola, Richard L. Brereton, Marleigh M. Sullivan, Nicholas K. Shonka, Tessa B. Wheeler, and William H. McDowell. "Novel ecosystems in the Anthropocene: a revision of the novel ecosystem concept for pragmatic applications." *Ecology and Society* 19, no. 2 (2014). DOI: http://dx.doi.org/10.5751/ES-06192-190212.

Norri, Juhani. *Dictionary of Medical Vocabulary in English, 1375–1550: Body Parts, Sicknesses, Instruments, and Medicinal Preparations.* New York: Routledge, 2016.

Nussbaum, Martha. *Hiding from Humanity: Disgust, Shame, and the Law.* Princeton, NJ: Princeton University Press, 2006.

O'Connor, Terry. "Commensal Species." In *The Oxford Handbook of Animal Studies*, edited by Linda Kalof. Oxford: Oxford University Press, 2014. DOI: 10.1093/oxfordhb/9780199927142.013.004.

*This present boke shewyth the manere of of hawkynge [and] huntynge and also of diuysynge of cote armours...* [The Book of Saint Albans]. Westminster, 1496. STC (2nd ed.) 3309.

*Treatyse of fysshynge wyth an Angle*. In *This present boke shewyth the manere of of hawkynge [and] huntynge and also of diuysynge of cote armours* ... [The Book of Saint Albans]. Westminster, 1496. STC (2nd ed.) 3309.

van Huis, Arnold, Joost Van Itterbeeck, Harmke Klunder, Esther Mertens, Afton Halloran, Giulia Muir, and Paul Vantomme. *Edible insects: future prospects for food and feed security*. FAO Forestry Paper 171. Rome: Food and Agriculture Organization of the United Nations, 2013.

Žižek, Slavoj, Eric L. Santner, and Kenneth Reinhard. *The Neighbor: Three Inquiries in Political Theology*. Chicago: University of Chicago Press, 2010.

# 2 Donne's Fish Biographies

It is easy enough to situate John Donne among early modern anglers. His biography was first written by Izaak Walton, whose *Compleat Angler* is the culmination of early modern angling literature and the subject of this book's final chapter. Walton and Donne had a friend in common: the diplomat and writer Henry Wotton, a man whose enthusiasm for freshwater fishing is documented in his own verse and in Walton's reminiscences. To illustrate how beneficial angling is for the mind, Walton speaks in the character of Piscator ("Fisherman") about the pleasure Wotton took in the sport:

> I do easily believe that peace, and patience, and a calm content did cohabit in the cheerful heart of Sir Henry Wotton, because I know, that when he was beyond seventy years of age he made this description of a part of the present pleasure that possest him, as he sate quietly in a Summers evening on a bank a Fishing.[1]

He goes on to quote (imperfectly) some piscatory lines from Wotton. As it was printed in *Reliquiae Wottonianae* in 1651, Wotton's angling poem begins:

> And now all Nature seem'd in Love,
> The lusty Sap began to move;
> New Juice did stir th' embracing Vines;
> And Birds had drawne their Valentines:
> The jealous Trout, that low did lie,
> Rose at a wel-dissembled Flie:
> There stood my friend, with patient Skill,
> Attending of his trembling quill.[2]

In his account of Wotton's life, Walton is sure to mention that the diplomat would not "forget his innate pleasure of Angling; which he would usually call, his idle time not idly spent, saying, he would rather live five May months then forty Decembers."[3]

Wotton was the recipient of some of Donne's most famous letters, in prose and verse; the two were close friends. Thus it is tempting to put Donne on the riverbank with Walton and Wotton, and that picture is very nearly realized in *The Compleat Angler*. In his jolly introduction to the poem that we now know as "The Bait," the character Viator confidently announces:

> I will speak you a Coppie of Verses that were made by Doctor Donne, and made to shew the world that hee could make soft and smooth Verses, when he thought them fit and worth his labour; and I love them the better, because they allude to Rivers, and fish, and fishing.[4]

Is Donne the "friend with patient skill" admired in Wotton's verse? Did he use a hazel rod? Did he tie his own flies? Can we now say for sure that the "trembling quill" is both a float and a pen? Alas, no. Donne's speaker ridicules the sport:

> Let others freeze with angling-reeds,
> And cut their legs with shells and weeds,
> Or treach'rously poor fish beset
> With strangling snare or windowy net.
> (17–20)[5]

"The Bait" responds to three poems. The first and second—"The Passionate Shepherd to his Love" by Christopher Marlowe and "The Nymph's Reply," usually ascribed to Walter Ralegh—both appear elsewhere in *The Compleat Angler*.[6] An anonymous third work is not included in Walton's text, despite the fact that it introduces piscatory recreation into the pastoral scene. In the ridiculously bountiful world of the anonymous reply, fish go "gliding on the sands, / Offering their bellies to your hands."[7] Donne builds on this image in particular, making a central conceit of fishing rather than listing it in a catalog of leisurely pursuits. The claim of "The Bait" is that the beloved, like Ovid's Arethusa, enchants all who see her as she bathes. She becomes an accidental angler, innocent of her own allure.

The fish, like the speaker, are fully aware that they will be somehow consumed by this creature, and they are willingly drawn in: "there th' enamoured fish will stay, / Begging themselves they may betray." ... But "thou need'st no such deceit," the speaker says, as artificial flies or nets, because "thou thyself art thy own bait:" / "That fish that is not caught thereby, / Alas, is wiser far than I" (25–8). The lyric plays on the familiar, cruel irony of angling: that the mechanism by which these animals become supper is their hungry pursuit of a meal for themselves. To figure erotic momentum as the appetite of a fish is to think about the mind of such a creature, if only in a cursory way. The speaker goes underwater and looks through "poor fishes' wand'ring eyes" at a massive, magnified, naked human body. The extreme difference between these two forms—the little piscine shape and the big human one—is one of the ways the poem gathers its humor. It's also a preoccupation that Donne carries with him well beyond his early erotic verse. If he chooses to champion indoor sport at the expense of angling in his witty volleys of the 1590s, Donne nevertheless returns to fish, again and again, always with an apparent desire to know what life feels like in their slippery, scaly skin.

Even if he was unlikely to go after them with rod and line, Donne openly admires the alien nature of fishes. Throughout his career, or from one career to the other, the poet-preacher uses fish to think through aspects of social and embodied life. Donnean fish defy generic boundaries, moving fluidly from epistle to elegy. They appear in his sermons, and not only when Donne speaks of the fishermen who became "fishers of men" in following Christ. Fish go swimming through the sermons, where they sometimes get caught and consumed, or have trouble becoming clean again when they "mud themselves" because, as Donne rightly explains, they "have no hands."[8] And they are nowhere more plentiful than in *Metempsychosis*, Donne's long narrative poem about the transmigration of souls, completed in the summer of 1601 and given pride of place at the start of the collected poems in 1633. While *Metempsychosis* has usually been read for its allegorical representations of political and religious difference,[9] the piscine lives in the poem are too vividly drawn to be collapsed into a wholly metaphorical reading. Without rejecting the possibilities presented by scholars less interested in fish, this study aims to open up the poem's deep stores of ichthyological knowledge. Reading *Metempsychosis* through the principles of cognitive ecology, we can observe how its fish live and die in an environment of minds.[10]

## Larval Life

What Donne exhibits, time and time again, is an interest in the edges of piscine life—the way a fish emerges into a swimming shape from its embryonic origins, the danger it faces of being captured and suddenly killed by any number of creatures. For a start, his attention is trained, in several places, on the texture and appearance of fish eggs. In the fresh waters of *Metempsychosis*, "a female fish's sandy roe" is "leavened" by the "jelly" of a passing male, resulting in a cluster of "small bodies" (222–25). In "An Elegy upon the Death of Mistress Bulstrode" (1609), the same sabulous texture is conjured, where "roes of living sand" are imagined as absorbing the water and turning it into land (15). The tremendous fertility of fish—the sheer quantity of living material that is produced by a single mating pair—elicited curiosity and comment from many early modern writers. In a 1600 treatise on fish and fruit that is largely given over to instructions on how to establish and maintain ponds, John

Taverner claims that "of all creatures fish are the greatest increasers in number;" if they weren't so keen on eating each other, the sea "and all fresh rivers likewise" would quickly be overpopulated.[11] His guide to pisciculture provides a lively account of fish reproduction:

> The ingendring and breeding of the like fish as aforesaid, I haue noted to be in this manner, sometime in May, and sometime in Iune, as the season happeneth to fall out apt for generation, the water by Gods prouidence hauing then a naturall warmth to performe the same, the male fish by course of nature, will chase about the female, seeking copulation: and as in all other creatures, so in this the female seemeth to shun and flie from the male, so that you shall see three, foure, or fiue male fish chase one female, and so hold her in on euerie side, that they will force her to swimme through weedes, grasse, rushes, straw, or any such like thing that is in the pond, wherein she being intangled and wearied with their chasing, they find oportunitie to ioyne in copulation with her, mingling their milt with her spawne, sometime one of them, sometime another, at which time the spawne falleth from her like little egges, and sticketh fast to the sayd weedes: some eight, nine, or ten dayes after which time it quickneth, taketh life, and hath the proportion of a fish: yea two or three dayes before it quicken, if you take such an egge and breake it vppon your naile, you shall perceiue the proportion of a fish therein. After it is quicke it mooueth very little for some fortnight or three weekes, and then it gathereth together into sculles by the shore side, where the water is shallow: howbeit the Tench frie will lie scattering in the weedes, and not flote in sculles.[12]

This system is so successful that the keeper of ponds must regulate the number of fry in his aquatic farm. It is best to let the pond dry out every other year, because "by that meanes you shall avoide superfluous number of frie, which greatly hinder the growth and goodnesse of your greater fish."[13] Paradoxically, one of the challenges of raising fish is that the creatures themselves are so good at reproducing. In a recent book about piscine perception and behavior, ethologist Jonathan Balcombe uses superlatives that confirm Taverner's assessment: the "fecundity" of fishes is "unmatched among vertebrates."[14]

Taverner, like Donne, at once personifies the fish and chronicles a life history that is radically non-human. Unlike us, with our direct development from four-limbed, big-headed infants to four-limbed, big-headed adults, fish undergo a series of relatively drastic changes on their way to sexual maturity. And since the fertilization of their eggs usually happens outside of the body, it is possible to witness the embryonic and larval stages of a piscine life cycle. Modern ichthyology textbooks describe the earliest stages in much the same way that Taverner does:

> The optic cups (the future eyes) and heart are the first organs to be identified easily. The heart starts to beat well before hatching and in some species the eyes become pigmented and functional, and the embryonic fish often wriggles and rotates within the chorion [egg case].[15]

And yet, although these developments may be relatively easy to see, they aren't necessarily easy to categorize. Those whose life work is the study of larvae would remind us that "stages exist in the mind of the biologist, not in the larva."[16] While human observers mark what seem to be the most significant structural changes in an animal, they don't always agree about what constitutes a new life stage. When Taverner and Donne envision the beginning of an individual piscine life, they consider the differences between embryonic, larval, and adult forms, trying to deliver an accurate account of processes that are gradual and alien.

Like modern biologists, these early moderns are cautious about naming the phases they see. Donne's *Metempsychosis* and Taverner's treatise both hesitate to give absolute answers—a newly hatched creature in Donne is "perchance a fish, but by no name you could call it" (230); Taverner admits of eels that "their manner of breeding is very uncertaine and unknowne."[17] As he gazes into the pond to watch the mating behavior of the fish and hovers above the eggs to watch for the first wiggle, Taverner incidentally narrates his own fascination with the perpetuation of piscine life. Similarly, Donne sets down a record of his curiosity about fish reproduction. At perhaps the very same time, the poet and the fish-farmer are writing about the moment when an individual life emerges and becomes a distinct self; both authors associate this quickening with anatomical structures and with the motility those structures make possible. In different ways, they write against the theory of spontaneous generation—Donne lets the movement of the soul proceed on its own course, separate from the cycles of sexual reproduction that bring new bodies into the world, while Taverner explicitly rejects the popular view that eels can emerge from dew alone. "I know that some hold opinion that they breede of the May deaw,"[18] he allows, but this belief is easily answered. The folk wisdom is to cut two blocks of fresh turf, set them in the water with their grassy sides together, and watch for baby eels in the grass the next day. There will be eels, Taverner says, but not because they are engendered by the dew. Use any dewless plant matter and the same result will follow. He explains, "The reason is, at that time of the yeare that river being fulle of such young Eeles, they will creepe into every thing that is sweete and pleasant."[19] It may be difficult or impossible to witness the sex that brings these beings into the world, but sex it is.

For Donne, too, the sparrow "o'erflutters his next hen," the fish "intertouch," and a wolf "engage[s]" a dog (194, 225, 421); new bodies emerge from the tender or violent intimacy between similar beings. In a narrative about the movement of a soul from one body to the next, it is surprising to find so many descriptions of copulation, a process which may seem to be beside the point. Certainly some readers have found the sexual episodes gratuitous. Herbert Grierson referred to the poem's frank, vivid sex scenes as "disgusting" and "wantonly repulsive" in his 1912 edition of the poems.[20]

But the attention to reproduction would be irrelevant and excessive only if Donne's questions were limited to the Pythagorean doctrine of transmigration. Instead, *Metempsychosis* is sometimes overwhelmingly concerned with bodily existence. Against the constant vitality and movement of the soul, the poet traces the limits that give meaning to that perpetual continuation: the first coalescence of shapes into a living form, and the final dissolution of that form that we call death. And beyond the rhetorical advantage of juxtaposing the immortality of the soul with the ephemeral nature of the body, the poem elicits wonder at the diversity and mystery of creaturely life by dwelling on—or *in*—non-human bodies. Because so much of *Metempsychosis* is given over to the experiences of fish, with fourteen of its fifty-two stanzas devoted to descriptions of fish and other aquatic creatures, it seems likely that Donne saw special potential in the comparison between human and piscine life histories. It isn't necessary, however, to read the poem's fish as convenient and fascinating points of contrast; the trope of transmigration could just as easily have been a handy opportunity for the poet to shift his shape, to take us with him into a series of animal frames, and to pose questions about other modes of being.

The first animal shapes that the soul of *Metempsychosis* inhabits are birds and fishes. It travels from the forbidden fruit in the Garden of Eden to a mandrake root that is pulled out of the ground by Eve, and settles in a sparrow's egg. The soul dwells in the body of a bird whose outrageous libido wears him out after three years, and then it makes its way into another kind of egg:

> The soul from her too-active organs fled
> T'a brook; a female fish's sandy roe
> With the male's jelly newly leavened was,
> For they had intertouched as they did pass,
> And one of those small bodies, fitted so,
> This soul informed, and abled it to row
> Itself with finny oars, which she did fit.
> Her scales seemed yet of parchment, and as yet
> Perchance a fish, but by no name you could call it.
>
> (221–30)

This body, like others in *Metempsychosis*, is "abled" by the soul that enters it. Although the creature's form has already been "fitted" through sexual reproduction, the individual is without a will until it has been "informed," at which point a second "fit" is made—the soul either provides the fins ("fit" in the sense of *supply* or *furnish*) or prepares them for use (makes them "fit" for the exertion to follow). The emergent selfhood of the hatchling is not yet a sure thing; it comes into its sexed identity subtly and incompletely, beginning as an "it," soon finding "her scales," and getting absorbed into the neuter pronoun once again. While the soul itself is always feminine, its hosts tend to alternate in gender throughout the poem; we expect a female fish after the male sparrow. Instead, the speaker hesitates to fix this "small body" in consistently gendered language. The imprecision is significant in a stanza that has everything to do with sex—the differentiation between male and female fish is essential to the beginning of this life story, but as the fish takes shape, the legibility of its not-quite-gendered body is less important than the will that propels it through the water with its newly animated "oars." The infant is fitted with motility before it can be made to fit into the categories of fish or female.

The speaker's taxonomic hesitation is reflected in ongoing scientific debates about larval forms and metamorphosis. The rhetorical diffidence of the observer seems to imitate the material vulnerability and in-betweenness of the hatchling. Piscine life at its earliest stages is like sand, jelly, and parchment: barely there, and therefore resistant to categorization. Donne's maybe-fish is larval, according to most biologists' definition: it is in a stage of its life history that is "morphologically distinct from the adult" and it will undergo indirect development. "Larvae must metamorphose to transform into, or be replaced by, adults."[21] As we've seen, however, definitions of *larva* and *metamorphosis* are unstable.[22] Some describe metamorphosis as "a dramatic and abrupt change in an animal's form or structure during postembryonic development."[23] But because there are different kinds of embryos, the event of hatching from the primary egg membrane is not always the most accurate marker for the beginning of larval existence; there are various definitions for "postembryonic."[24] Further, it can be difficult to say how drastic a change needs to be in order to qualify as metamorphosis. Ichthyologists have not reached a consensus about how and when to apply the term to the early life stages of fish.[25] Such stages are "given a variety of names, the meaning of which is often rather imprecise."[26] So far, the uncertainty of Donne's speaker makes perfect sense.

An alevin, as recently hatched trout and salmon are now called, swims along with its yolk sac attached to its body. The hatchling is considerably different from an adult in both shape and behavior; it will absorb its yolk, develop scales, change color, and grow much larger as it reaches maturity.[27] Its "embryonic, undifferentiated fins" will emerge and gradually take shapes and positions appropriate to their functions.[28] The larva may well have "the proportion of a fish," to use Taverner's expression, since *proportion* can mean either that it is a small likeness of a grown-up or that it is a mere fraction of what it will become.[29] Its chances of survival are slender; only 5.3 percent of the "small bodies" Donne imagines in a cohort of freshwater fish eggs will make it through the larval stage.[30] If it manages to eat an equivalent of half its body weight every day and escape predation, Donne's infant creature will eventually be a fish, but its status as it first moves from embryo to open water is much less certain.[31]

## Digestion

The maybe-fish does not number in the lucky 5 percent. It is immediately consumed by a passing swan whose robust, decidedly adult body is contrasted with the tiny alevin to pathetic effect: "goodly, like a ship in her full trim, / A swan … / Glided along … / It moved with state, as if to look upon / Low things it scorned" (231–37). The unseen predator catches "the poor fish" and swallows "much such" with it, devouring any creature that is not "too swift, too great, or well-armed" (235, 239–40). Even if the alevin had seen its hunter, its "finny oars" would be too feeble to motor it out of the path of the ship-like swan. It is transported instead into the digestive tract of the ravenous bird, where Donne's speaker follows it. Moving quickly from the beginning of a piscine life to its end, the episode now turns to a brief contemplation of what happens to a body that has been eaten. The swallowing of the fish does not instantly bring about its death or the departure of its soul; instead, the soul is temporarily trapped in a double enclosure:

> Now swam a prison in a prison put,
> And now this soul in double walls was shut
> Till, melted with the swan's digestive fire,
> She left her house the fish, and vapoured forth.
> (241–44)

The metaphor of incarceration is borrowed not only from the long tradition of construing the earthly body as a prison for the soul, but also from popular conceptions of fish stuck in landlocked waters or intentionally kept in ponds. The *Treatyse of fysshynge wyth an Angle* says that a pool of standing water is "but a prison to fysshe." Gervase Markham, in his instructions for aquaculture, draws a sharp distinction between "the Fish in Riuers [who] haue, by vertue of the current, ever something brought to them to feed on" and "the Fish which is imprisoned in Ponds and wants that helpe."[32] Because fish are so mobile by nature, and because the art of farming them was gaining new popularity, the image of a captive piscine body would have been poignantly familiar at the moment of Donne's writing. Added to that association is the perfect irony of the "digestive fire" that removes the infant fish permanently from its home element of water, causing its soul to liquefy and evaporate. As it chases the soul from the body, this heat presumably brings physical death to the tiny, transparent alevin.

From his sermons, it is clear that Donne was fascinated by the question of how one body digests another, and that he would continue, many years after the composition of *Metempsychosis*, to examine the boundaries of life and death through the lens of piscine experience. Preaching at Lincoln's Inn in May 1620 on the doctrine of resurrection, Donne explains Job's certainty that he will "see God ... in [his] flesh" even after the decomposition of his corpse.[33] His sources for the sermon are church fathers Justin Martyr and Athenagoras, both of whom insist that "we expect to receive again our own bodies, though they be dead and buried in the earth."[34] To Justin's relatively abstract statement that "it is not impossible for the bodies of men and women, dissolved and like seeds resolved into the earth, to rise again in God's appointed time and put on incorruption,"[35] Athenagoras adds a vivid thought experiment. This is how fish come swimming into Donne's sermon:

> Such men [those who deny the doctrine] say, for example, that many human bodies become, through drowning by shipwreck or in rivers, food for fishes; many of those who die in war or by some other hard fate or whim of circumstance lie unburied, are food for the first animals that approach. Bodies then, they say, are thus done away and their proper portions and particles pass after decay into a multitude of animals and are by digestion united to the bodies that thus feed upon them, so that the separating out of these becomes impossible.[36]

Opponents to the doctrine would argue that this essential separation of parts is even more difficult when a human being eats an animal who has eaten human flesh. To this "further and more awkward predicament" Athenagoras replies that "what is unnatural never turns into nourishment for the parts and portions that need it," so "human bodies can never be incorporated into other human bodies that use them as unnatural food, even should they by some unhappy fate pass through the bowels of such men."[37]

Like Justin and Athenagoras, Donne uses God's power of creation as an argument in favor of the doctrine. The sermon says that humans can be made whole again no matter what, since God was able in the beginning to make everything out of nothing:

> The specification of creatures, and the disposing of them, into their severall kinds, the making of that which was made something of nothing before, a particular thing, a beast, a fowl, a fish, a plant, a man, a Sun or Moon, was not so hard, as the first production out of nothing.
>
> (6)

The fish that is absent from Justin and mentioned only in passing by Athenagoras will become Donne's chief example of "specification," and thus the most compelling vehicle for his meditation on the consumption and eventual resurrection of human bodies. Of all creatures, fish are perhaps the easiest to imagine as both feeding and feeding *upon* humans, so they provide a convenient topos for both the initial problem of us-as-food and the "predicament" of humans incorporating the flesh of fellow humans in a looped food chain. Donne streamlines Athenagoras's refutation:

> Does this trouble thee … that one man is devoured by a fish, and then another man that eats the flesh of that fish, eats, and becomes that other man? … that first man did not become that fish that eate him, nor did that fish become that second man, that eate it; … both that man, and that fish are resolved into their own elements, of which they were made at first.
>
> (6–7)

The extreme difference between human and piscine forms, exemplified by the fact that we cannot survive long in each other's home environment, is precisely what makes this illustration so apt. Donne exploits the contrast in a later sermon, delivered during Lent in 1623, where human sin is figured as a metamorphic process. The sinner "hath passed many transformations, from shape to shape, from sin to sin, (he hath been a Salamander and lived in the fire … and then he hath beene a Serpent, a Fish, and lived in the waters, in the water successively, in the troubled water of sedition in his youth, and in his age in the cold waters of indevotion)."[38] These metaphors are built on the deep contrast between human morphology and the bodies of other creatures—slippery denizens of habitats that would prove fatal to us. Sin breaks a person up much more severely than a hungry wolf or fish might, so simple bodily resurrection is not the most difficult task; it's gathering a person back together after he has been "resolv'd into diverse substances, scattered into diverse forms of severall sinnes" that is "the greatest worke."[39] The sinner, worse than having been eaten and digested, has adopted alien ways of living, has so radically compromised his humanity that he now belongs in fire and water. It's telling that Donne shifts from "serpent" to "fish" in order to exploit this elemental difference; natural histories of the sixteenth and seventeenth centuries mark a distinction between serpents and fish, but the preacher wants to conjure a decidedly aquatic creature that is *both*, something that embodies at once the spiritual dread of "the serpent" and the real-world weirdness of a fish.[40]

In the earlier sermon, that weirdness is largely attributable to the shifting role of fish in our trophic web. Fish and human beings represent each other's demise; eating each other is one of the few ways in which we interact. The costly intimacy of such consumption either depends upon or brings about the death of one party: the fish is caught, killed, and eaten; the human being is served up by drowning. For an audience familiar with that equation, and curious about the fate of flesh that has entered an alien digestive system, there is perhaps no better way to describe the mystery of bodily resurrection. In Donne's formulation from the Lincoln's Inn sermon on Job, humans become fish food, "scattered" into innumerable parts, but this metabolic tangle is easily undone: "I, the same body, and the same soul, shall be recompact again, and be identically, numerically, individually the same man ... I shall be all there" (19).

Donne juxtaposes the miracle of such a reconstruction against the more familiar process of decomposition. From the story of Job come the hungry earthworms that figure so prominently in the sermon as decomposers: "And though, after my skin, wormes destroy this body, yet in my flesh shall I see God" (1). Donne explains that the worms are present in some familiar translations of Job's language but "not in the Originall," and yet they are an admissible embellishment because they make "the destruction the more contemptible" (16). The annelids that have wormed their way into the narrative give the preacher a precedent for the especially arresting claim that "this flesh which I have now, is not *mine*, but the wormes" (3). They do not, however, offer a quotidian illustration of the human being as both consumed and consumer; for that, the preacher needs fish. Could Donne have been thinking of Hamlet's mocking demonstration of "how a king may go a progress through the guts of a beggar" (4.3.29–30)?: "A man may fish with the worm that hath eat of a king and eat of the fish that hath fed of that worm" (4.3.26–27).[41] In both the tragedy and the sermon, the image of feasting annelids, effective as it may be at conjuring disgust, fails on its own to demonstrate how human beings might unwittingly consume parts of their conspecifics. The riddling prince and the poet-preacher exploit the wriggly repulsiveness of worms, but it's fish that close the circle. The threat of indirect, belated cannibalism is exposed through the appetites of fish. Piscine hunger, oriented as it is toward both annelid and human flesh, also invites audiences to recognize themselves *in* worms. In this way, the sermon on resurrection surely continues the work of *Metempsychosis*, bringing human embodiment into comparison with neighboring but fundamentally alien forms.

The sermon also reflects *Metempsychosis*'s imagery of the vulnerable membrane that separates a body from its environment. The not-quite-scales on the skin of the alevin are like "parchment" (229); the simile suggests that the newly hatched fish is an unwritten story, an incipient text. In the sermon, the fragility of human skin is contrasted with the more stable material inside of it: "wee have the character, and seal of God stamped in us, in our baptism; and, all this is bound up in this velim, in this parchmin, in this skin of ours, and we neglect Book, and image, and character, and seal, and all for the covering" (13–14). Parchment is used, then, to describe the vulnerability of the body at both extremes of life. Donne's "velim," figured as a binding for the divinely stamped human life, is an expanse of skin named for the calf from whose body it has been removed (and thus derived from the Latin *vitellus*), but it bears a striking resemblance to the word biologists now apply to "a membrane or membranous integument": *velum*, from the Latin meaning "covering."[42] According to the sermon's account of how "Corruption seises the skinne, all outward beauty quickly," we would expect the delicate outer layer of the alevin to perish first in the "digestive fire" of the swan's gut. Unlike the confident preacher, however, the infant fish will not be "recompact again," at least not within the timeline of the poem. The soul does not wait for any such reunion, but makes the most horizontal move to be found in all of *Metempsychosis*: "Fate, not affording bodies of more worth / For her as yet, bids her again retire / T'another fish" (245–47).

At this point in the poem's sequence of episodes, we have seen the soul travel from one plant to another—the Edenic apple to the mandrake—and later it will go from parent to offspring as it inhabits first a wolf and then its half-wolf, half-dog progeny. Nowhere else in the series of transmigrations does the soul make such a perfectly lateral transition as this, from fish to fish. The slender differentiation between the two fish marks the enormous distance to be covered before the soul reaches the strange, large, human body that is said to be its present home. In the final movement of the soul, the poet illuminates not its ultimate residence, but the body of Themech—a daughter of Adam and Eve who marries her brother Cain. Presumably, many human hosts follow; the beginning of the poem indicates that the soul has been in the bodies of Mahomet and Luther, "and lived where every great change did come" (66, 69). But these migrations are not narrated. It would seem that the diversity of piscine life was sometimes more interesting to Donne than the trumped-up differences between members of his own species.

## Silence and Breath

In the stanzas between its hatching at its death, poem's second fish survives the stage of life during which the first perished, and seems to be maturing into an adult salmon. But neither the species nor its various life stages are named; in the analysis that follows, the word *fry* preserves Donne's ambiguity and allows us to think of the fish as a representative youngster. Derived from an Old Norse word for seed, *fry* was the appellation for roe or infant fish in Middle and early modern English, but it could just as easily refer to the offspring of any species.[43] It suggests the littleness and vulnerability of the new fish, without yet specifying a single phase of its development. (Later, this second fish does show signs of being at a very particular moment in its life history.) It also chimes with Taverner's usage in his pond-building treatise, where he marvels not only at the great numbers of fry produced by adult fish but also at the fantastic appetite one grown fish might have for the offspring of another: "I have seene in the belly of a small Perch sixteene or seventeene small Carpe frie at once."[44] Taken as a collective, fry are plentiful; as individuals, they are likely to be eaten.

Although the second fry lasts for twice as many lines as the first, its mortality is introduced immediately. The small aquatic body is "to any new desire / Made a new prey" (247–8); it is subject to the violent whims of whatever creature might behold it. There is only a moment of free-swimming life before the fish is defined by its surroundings—the unfamiliar environment of minds into which it has emerged—as prey. This vulnerability is attributable to the fish's inability to fight back, a deficit most poignantly figured in the animal's silence. Comparing the human faculty of speech to the voicelessness of the fish, the speaker here provides an epigrammatic statement that takes us out of the water: "he that can to none / Resistance make, nor complaint, sure is gone. / Weakness invites, but silence feasts oppressïon" (248–50). In this paraphrase of the proverb "Silence is consent,"[45] Donne seems to reduce the fish to a political metaphor. Here and elsewhere, he sketches out the contours of human speech—its urgency, its power, and its dangers—by juxtaposing it against piscine quietness.

In a famous verse epistle to Henry Wotton ("Sir, more than kisses, letters mingle souls"), likely written in the summer of 1598, Donne's advice is that his friend should be at ease within himself and take the middle way. For that first idea, "Be thine own palace, or the world's thy jail," Donne can rely on the proverbial snail who is, naturally, always at home. For the other notion, that one can accomplish most by moving subtly through the social world, the poet uses fish:

> And in the world's sea, do not, like cork, sleep
> Upon the waters' face; nor in the deep
> Sink like a lead without a line: but as
> Fishes glide, leaving no print where they pass,
> Nor making sound, so, closely thy course go;
> Let men dispute whether thou breathe or no.
> (53–8)

Up to this point, the dominant conceit of the poem has been a journey by sea—"Life is a voyage" (7)—during which one's ship might be damaged or held up by various misfortunes. The metaphor seems to continue here, but there is a shift from "the world's sea" to the familiar riverside position of an angler. The fish in this riparian image are not the threatening remoras or torpedoes of the open ocean (8, 18), but freshwater neighbors whose behavior Wotton would recognize. Don't be like the tackle, Donne says; be like the fish, the living body that can move undetected from place to place. The silence that will prove fatal to the second fish in *Metempsychosis* is an advantage here; like its ability to travel without making any disturbance (or, in the terms of the later poem, showing any "resistance"), the noiselessness of the fish protects it from predation.

A later transmigration in *Metempsychosis* will find the soul in the body of a whale who is threatened by "two small fishes," assassins whose conspiracy is protected because "they fishes were, and could not speak" (348). Just as violence is surprisingly quiet in that episode, the impressive fecundity of fish is the more wonderful for their "monastic silence" in Donne's elegy on the death of Cecilia Bulstrode (14). These "harmless fish" produce offspring at a death-defying rate, but make no sound while they're at it. Whenever he speaks of speechless fish, Donne picks up a thread that stretches back to the classical period and was commonly woven into early modern texts.[46] Erasmus was especially fond of the idea (and, as it happens, particularly averse to fish).[47] Two of his adages deal centrally with piscine silence, and in one of his colloquies, he has a butcher declare that fish-eaters resemble the creatures they eat: "Generally men full of fish are themselves like fish: pale, stinking, stupid, silent."[48] For Donne, the silence of fish persists in spite of their remarkable capacity for intentional action. If it looks at times like powerlessness, piscine muteness elsewhere appears to be strategic and advantageous.

The second fry of *Metempsychosis* does seem to have taken the advice of "Sir, more than kisses." It goes with the flow: "Pace with the native stream this fish doth keep, / And journeys with her t'ward the glassy deep" (251–2). On its way, the fish is nearly caught in two different ways, but its size allows it to escape. It has the slow progress of fishing technology to thank:

> Once with a hidden net
> Though with great windows (for when need first taught
> These tricks to catch food, then they were not
> As now, with curious greediness to let
> None scape, but few, and fit for use, to get),
> As in this trap a ravenous pike was ta'en,
> Who, though himself distressed, would fain have slain
> This wretch (so hardly are ill habits left again).
>
> (254–60)

Two comparable vehicles of desire threaten the little fish: the enterprise of human beings and the stubborn habits of the "ravenous pike." As in the previous episode, the fish is yet too young to be assigned a species, while its potential captors are identified more precisely. The windowy net that would catch the fry is representative not of humankind in general, but of an earlier stage in piscatory history when people had neither the ingenuity nor the "curious greediness" to fashion tighter webs. Motivated by "need" to catch only those fish that were "fit for use," the fishers of the past contrived no technology that would, like the stately swan, consume everything in its path. The sweeping net appears only at first to be a threat; it quickly entraps the predatory pike and lets the young fish swim away unscathed. Instead of presenting the primitive net as deficient, the speaker suggests that openness is an intentional feature of its design. Since that more humane time, people have become pike-like, letting the "ill habits" of the hunt take over in spite of their own best interests.

After the account of innovations in fishing gear has drawn human greed into comparison with piscine hunger, the text provides further reasons to wonder how similar fish and people may be—not just in terms of shared desires, but on the level of physiology. As the fry swims through the net and toward safety, the speaker tries to see inside its body, and considers whether the animal's breath is produced through a distilling apparatus like the one used by alchemists. He looks for tiny alembics ("limbecks"):

> … whether she leap up sometimes to breathe
> And suck in air, or find it underneath,
> Or working parts like limbecks hath
> To make the water thin and air-like, faith
> Cares not …
>
> (264–8)

The question of whether fish breathe the same air we do, and the subsequent question of *how* they might do so, was something that fascinated Donne. It comes up, of course, in his letter to Wotton, where the quietness of fish is an example for any wily secretary to follow: "Let men dispute whether thou breathe or no" (58).

As Robin Robbins observes, the alchemical mystery surrounding the silent fish's respiratory system in the verse epistle is like the imperceptible boundary between life and death in the ornate opening simile of "A Valediction Forbidding Mourning," a poem composed some years later:[49]

> As virtuous men pass mildly away,
> And whisper to their souls to go,
> And some of their said friends do say,
> 'The breath goes now', and some say, 'No,'
> So let us melt and make no noise …
>
> (1–5)

It is as hard to observe the breath of a fish as it is to detect that of a dying man—if, that is, the man's virtue allows him to die peacefully. Wotton, according to his friend's advice, should "pass mildly" through the various political elements of country, city, and court, knowing himself but remaining mysterious to others. Judging by the account provided in Izaak Walton's biography of Donne, the poet maintained just the kind of quiet countenance he urges Wotton to display: "His aspect was cheerful, and such as gave a silent testimony of a clear knowing soul, and of a Conscience at peace with itself."[50] At the moment of the soul's departure, a peaceful conscience would be fishlike, and "make no noise." Read in tandem, Donne's two similes establish an association between two unknowable kinds of vitality: the life that comes after death, and the life that belongs to fish.

The second piscine episode of *Metempsychosis* introduces a series of hypotheses about fish respiration before dismissing the matter altogether on the grounds that "faith / Cares not." The dismissal comes too late to be convincing, given the momentum of the three fantastical illustrations that invite us to imagine fish filling their lungs in different ways. Here, as elsewhere in Donne, the respiratory system that cannot be monitored is a question that remains open. In the case of *Metempsychosis*, it is a secret the fish keeps. The reader can pause, however, where the poet does not, if only to note that the hypothetical systems passingly mentioned in the poem are, like so many of Donne's guesses about life as a fish, fairly accurate. Most fish do find their air "underneath," in the water itself. Gills are the "working parts" usually responsible for deriving oxygen from the water that passes over them. And many freshwater species do breathe air, although the structures they use to do so (the scales, the hindgut) are sometimes surprising.[51] Having introduced the intriguing riddle of respiration, the speaker quickly changes tack, calling himself back from tangential speculation only to venture further downstream with the largely unknowable fish.

## Anadromy

Although the fry is given no precise name, its movement outward from stream to sea suggests that it belongs to an anadromous population—a group of fish that begin life in fresh water, move out to the ocean as juveniles, and swim upstream to spawn when they reach maturity. The fish in question may be a salmon, a shad, a sturgeon, or a smelt, to list just a few of the possibilities. What is certain is that the youngster exhibits behavior that is typical of anadromous fish on their first encounter with salt water. After insisting that breathing mechanisms are irrelevant to the present narrative ("faith / Cares not"), the speaker takes a moment to situate the fry at a confluence of waters and to describe this mixed environment in terms of its relative safety for the fish: "but safe the place she's come unto, / Where fresh with salt waves meet, and what to do / She knows not, but between both makes a board or two" (268–70). The confusion of the little fish, evident in her movements to and fro, makes the estuary legible to a land-bound spectator. Her behavior marks the water's brackishness. Tacking back and forth, hesitating over the decision to continue the swim seaward, this fish acts exactly as cognitive ecologists would expect it to.

Fish have to regulate the amounts of salt and water in their bodies through systems of absorption and donation that differ according to their habitats. Depending on where it swims, a fish's internal system may be salty or less salty than the environment immediately surrounding it: "The concentration of salts maintained internally by a typical fish is higher than that of freshwater and lower than that of saltwater."[52] To help set the balance, ocean fish have specialized chloride cells in their gills, and several ways of giving excess salt back to their environment: they "drink copiously, absorb large amounts of seawater across the lining of the intestine, and secrete only small amounts of urine that are relatively high in salt concentrations."[53] Conversely, fish in fresh water have to donate water in order to survive; they drink less but produce more urine. Because of the different exigencies of living in different aquatic environments, anadromous fish make drastic internal changes to their bodies. For example, "When they grow from the freshwater parr stage to the smolt stage and head downstream to the ocean, young salmon and sea-going trout must change physiologically from hoarding salts and discarding water to the opposite strategy."[54]

This shift is sometimes accompanied by new behavior; one researcher working with radio-tagged fish "found that during the downstream run, wild Atlantic salmon smolts hesitated upon encountering saltwater, often turning back or remaining in place for some time before continuing on out into the ocean."[55] Like Donne's fry, these young fish were visibly uncertain about travelling out to sea. If the second fish of *Metempsychosis* can be read as an Atlantic salmon (*Salmo salar*), the poem narrates the process of smolting: the seaward migration after which this particular life stage is named. Donne employs neither the old verb *to smolt*—meaning "to make off, go, escape" or "to go free"[56]—nor the noun that had, for over a century, been the standard name for a young salmon making its way to sea,[57] but the episode delivers a strikingly precise description of typical smolt behavior.

The puzzling predicament in which the fry finds herself is literally magnified as the reader's perspective is lifted to achieve an aerial view into the clear water. Opposite to the fish's uncertainty is her enhanced visibility under this aquatic lens—an ironic failure of hospitality: "So far from hiding her guests water is, / That she shows them in bigger quantities / Than they are" (271–3). Such magnification is an implied effect of the water in "The Bait," when the speaker of that poem compares the beloved's body to the sun and moon, or teases her about her reluctance "to be so seen" (13). There, the human body is already humorously massive in relation to its piscine pursuers, a giant bait that looks even bigger under water. In *Metempsychosis*, the first two fish are tiny compared to their predators, but the second fry is made visible (and therefore "to any new desire / Made a new prey" [247–8]) by the water that is supposed to be her home element:

> Thus, doubtful of her way,
> For game and not for hunger, a sea-pie
> Spied through this trait'rous spectacle from high
> The seely fish where it disputing lay;
> And, t'end her doubts and her, bears her away.
> (273–6)

The piscine form, in ways that Donne will continue to emphasize, is always in danger. Hesitating with the fry at this estuarial juncture, the episode shows not only that ordinary life is perilous for fish, but that the piscine mind labors to interpret the dangers in its environment. The fry "knows not;" she is "doubtful" and "disputing" (270, 273, 276). As she is about to be devoured, the fish is also "seely"; this word, antecedent to the modern English *silly*, signals both the mental limitation and the pitiable vulnerability of the developing young animal. In theological thought reaching back to Donne's ancestor Thomas More and beyond, the human soul is often described as *sely* (the more common spelling) in relation to God. More wrote of "the paynys that sely sowlys feele when they be departed hence."[58] Like the captured fish, the soul is subject to fierce and unpredictable judgment. The fry itself is doubly silly, then, as its soul enters and leaves its vulnerable little body.

But the seeming silliness of the fish is only one piece of a larger cognitive picture. Her unfamiliarity with salt water—the fact that she "knows not" what to do at this confluence—is not a representative ignorance but an impediment to careful decision-making, important for the breach it marks in the fish's ability, up to this point, to read her surroundings and choose a way forward. As her whole body is busy processing information, the fish's motions and pauses invite the attention of another mind. She is endangered by her own efforts to select a safe passage, inadvertently sending the sea-pie a clear signal from the water below. There is irony in the stark visibility of this bird (*Haematopus ostralegus*), which is now more commonly called an oystercatcher but was known in early modern taxonomy for its vividly contrasting black and white plumage. With the striking pied body come dashes of sunset: bubblegum-pink legs, an orange-vermilion beak, and scarlet eyes.

Unsurprisingly, at the time of Donne's writing, the sea-pie's appetite for fish had not gone unnoticed. Leonard Mascall, in his 1590 angling manual, documents the sportfisher's need to protect fish from the appetites of "such deuourers and rave[n]ers as hath and will destroy them;"[59] to that end, he provides special instructions for the capture of sea-pies. Like their prey, oystercatchers move between sea and fresh water. For them, the estuary is a place where one can take advantage of hesitant young fish:

> The Sea-pie is a foule that useth the seas, and breedes much in Ilands in the sea, and liues most by fish and wormes, and where as they use in fresh riuers, they destroy much fish, young frie, and such as swimmes nigh the toppe of the water, and will be in shallow places of the water: and there they haunt to take and feede on them. Therefore the fisher men haue inuented a way howe to take them ...[60]

The trick is to attach a cross of limed osier twigs to the body of a bait fish and place it in a visible spot on the water's surface, making it seem alive and available (like the fry in *Metempsychosis*). Sea-pies will swoop down and take such bait "as soon as they shall see it[;] they will take and flie away with it in their bylles, and soone they shall be lymed therewith."[61] But in the absence of a crafty angler, the fry of Donne's poem has little chance of escaping the long, fiery bill of the oystercatcher.

The cognitive content of the predator's behavior is also part of the scene: the oystercatcher's "new desire" is not for food but for sport. Like many anglers depicted in early modern texts, this creature fishes for fun rather than for sustenance. The gratuitousness of the capture

*Figure 2.1* Baited devices for capturing sea-pies, from Leonard Mascall's *Booke of fishing with Hooke & Line*, 1590. Houghton Library, Harvard University.

is pointed up by the access we are granted to the mind of the prey. The fish is all cognition; she is coterminous with her own "doubts." The "way" that she strives to find is abruptly subsumed in the rhyme that permanently "bears her away," displacing her thoughts with the less sophisticated but more aggressive will of the oystercatcher. While Mascall encourages the culling of sea-pies as competitors, Donne insists that our species is no better. He stocks his verse with plenty of fish, but the quantity and diversity of piscine life is always threatened by the range of swift deaths that might descend upon these individuals. Without dissolving into pure metaphor, fish become emblematic of both fecundity and mortality.

## Violent Ends

The portrayal of piscine death having reached a certain pitch, the narrative is now suspended to make way for a sustained rhetorical question: "Is any kind subject to rape like fish?" (281). The indignant (or mocking) speaker notices that although they do not harass human beings, hunt beasts, chase birds, or interfere with the offspring of any animal, fish are killed by all of us: "Yet them all these unkind kinds feed upon" (288). Humans are the final and worst predators in the poem's catalogue of violent fish-eaters, with our unnecessary sports and arbitrary edicts about what to eat when: "To kill them is an occupatïon, / And laws make fasts and Lents for their destructïon" (289–90). Religious prescriptions are on a level with leisure activities in this formulation, and both are implicitly compared to the conduct of the unhungry sea-pie in the previous stanza. There is no observable *need* to capture and consume fish, and yet the practice persists among many species with many different apparent motivations.

The gobbling of the second fish is a further opportunity for Donne to consider what happens when one creature is consumed by another—not in the distinctly theological way in which he takes up the question in the sermons, but nevertheless with a sense of wonder about where one being ends and another begins. The oystercatcher is blown further seaward by a great wind, and flies until he is exhausted and stranded, then dies. The fish in his belly lives until this moment and expires at the same instant: "The souls, no longer foes, two ways did err" (298). Continuing to differentiate between the two identities, the speaker holds them in their most recent bodies in order to confirm that his narrative course remains with the soul that was introduced at the beginning of the poem: "The fish I follow, and keep no calendar / Of t'other; he lives yet in some great officer" (299–300). This declaration invites allegorical reading, but that's not all—it is interesting that the speaker should refer to the soul that has just departed from the fry's body as "the fish," as though there is something piscine about the soul itself. After inhabiting two little fry, the soul has absorbed a fishy identity. Once again, it can move only into another aquatic form.

The next migration looks, at first, to be just as lateral as the last: "Into an embryon fish our soul is thrown" (301). But the next home for the poem's protagonist is the body of a whale, more precisely named some stanzas later (324) and huge in comparison with the tiny forms that have come before it. As we've seen, Donne's liberal usage of *fish* is not unusual; early modern people could be aware of the difference between gilled, finned fishes, and huge marine mammals and nevertheless use the term *fish* to refer to any aquatic creature (as many people do today). Conversely, *whale* came with its own imprecision; the word was applied to the freshwater species *Silurus glanis* (the sheat-fish) because of its massive size, and *fish* and *whale* were used interchangeably in reference to the story of Jonah.[62] The whale of *Metempsychosis* will be killed by two creatures—"The flail-finned thresher and steel-beaked sword-fish" (351)—whose allegorical identities seem to be more important than the species to which they belong. Notwithstanding, these animals—perhaps the fox-shark *Alopias vulpes* and the sharply named *Xiphias gladius*—bring the narration back to piscine forms, marking a contrast between "the sluggard whale" (354) and the agility of its non-mammalian killers.

For the first time since the brief encounter with a "ravenous pike," the poem offers up an image of violent fish. These large, oceanic species have little in common with the delicate riparian forms that first carry the soul through the water, but they serve as a reminder of the tremendous diversity and illegibility of fishes. In their mute conspiracy, the thresher and swordfish show that the absence of human speech is not necessarily tantamount to an inability to communicate, and that silence might, anyway, be a function of our hearing rather than a characteristic of fish. Making killers of these final fish, Donne gives violence back to the creatures who are most "subject to rape." Maybe he is taking a jab of his own at the self-satisfaction of his sportfishing friends, or maybe all of the piscine forms introduced in *Metempsychosis* are merely representatives of particular people in the poet's world. One thing is clear: although Donne may never have caught a fish, they certainly captured him.

## Notes

1 Izaak Walton, *The Compleat Angler* (1653, Wing [2nd ed.] W661), 33–34. Subsequent references in this chapter are to this, the first edition of the *Angler*.

2 Henry Wotton, "On a Banck as I sate a Fishing, A Description of the Spring" in *Reliquiae Wottonianae* (1651, Wing [2nd ed.] W3648), 524 (1–8). The *quill* mentioned here is a hollow length of the shaft of a feather, used as a float in angling; the available pun on writing—an art that relies on another kind of "trembling quill"—goes unexploited.

3 Izaak Walton, "The Life of Sir Henry Wotton," preface to *Reliquiae Wottonianae*.

4 Walton, *The Compleat Angler*, 184.
5 John Donne, "The Bait," in *The Complete Poems of John Donne*, ed. Robin Robbins (Harlow, UK: Pearson Education, 2010). Text citations to Donne's poetry refer to line numbers in this edition.
6 Walton, *The Compleat Angler*, 66–68.
7 Quoted in *The Complete Poems of John Donne*, ed. Robin Robbins, 133. All three poems appear in sequence in *Englands Helicon* (1600, STC [2nd ed.] 3191), Aa1v-Aa3v.
8 John Donne, "Untitled [second sermon preached at the Hague]" in *The Sermons of John Donne*, ed. Evelyn M. Simpson and George R. Potter (Berkeley and Los Angeles: University of California Press, 1955), vol. 2, sermon 14, 20, http://contentdm.lib.byu.edu/cdm/compoundobject/collection/JohnDonne/id/3146/rec/2.
9 The most thorough recent study of the poem is Siobhán Collins's *Bodies, Politics and Transformations: John Donne's Metempsychosis* (Farnham and Burlington: Ashgate, 2013), which includes a chapter devoted to birds and fish. Collins contends that "*Metempsychosis* is a deeply philosophical text, concerned with temporality, embodiment and transformation, and thus self-consciously fragmented, open and equivocal in form; and that it is the most major and vital of Donne's poems for its remarkable openness to and its extraordinarily energetic and inventive portrayal of conflict within identity" (9).
10 I borrow this phrase from Eleanor Rosch's chapter "The Environment of Minds: Toward a Noetic and Hedonic Ecology," where she argues that the "dualism in which mind and environment are seen as separate objects that can be studied as such ... may be a metaphysic of our culture rather than an incontrovertible fact of nature." In *Cognitive Ecology*, ed. Morton P. Friedman and Edward C. Carterette (San Diego, CA: Academic Press, 1996), 22.
11 John Taverner, *Certaine experiments concerning fish and fruite: practised by Iohn Tauerner Gentleman, and by him published for the benefit of others* (1600), STC (2nd ed.) 23708, 17.
12 Taverner, *Certaine experiments*, 18.
13 Taverner, *Certaine experiments*, 12.
14 Jonathan Balcombe, *What a Fish Knows: The Inner Lives of Our Underwater Cousins* (New York: Scientific American/Farrar, Straus and Giroux, 2016), 17.
15 Quentin Bone and Richard H. Moore, *Biology of Fishes*, 3rd ed. (New York and Abingdon: Taylor & Francis Group, 2008), 230.
16 Carole S. Hickman, "Larvae in Invertebrate Development and Evolution," in *The Origin and Evolution of Larval Forms*, ed. Brian K. Hall and Marvalee H. Wake (San Diego, CA: Academic Press, 1999), 27.
17 Taverner, *Certaine experiments*, 19.
18 Taverner, *Certaine experiments*, 19.
19 Taverner, *Certaine experiments*, 20.
20 Herbert J. C. Grierson, "Donne's Love Poetry," in *John Donne: A Collection of Critical Essays*, ed. Helen Gardner (Englewood Cliffs, NJ: Prentice Hall, 1962), 33, quoted in Collins, *Bodies, Politics and Transformations*, 155.
21 Brian K. Hall and Marvalee H. Wake, "Introduction: Larval Development, Evolution, and Ecology," in *The Origin and Evolution of Larval Forms*, ed. Hall and Wake, 1.
22 Hall and Wake, "Introduction," 2.

23 S. Laurie Sanderson and Sarah J. Kupferberg, "Development and Evolution of Aquatic Larval Feeding Mechanisms," in *The Origin and Evolution of Larval Forms*, ed. Hall and Wake, 305.
24 Hickman, "Larvae in Invertebrate Development and Evolution," 25–26.
25 Sanderson and Kupferberg, "Development and Evolution," 305.
26 Peter S. Maitland, *The Hamlyn Guide to Freshwater Fishes of Britain and Europe* (London: Hamlyn, 1977), 24.
27 Sanderson and Kupferberg, "Development and Evolution," 306.
28 Maitland, *The Hamlyn Guide*, 24; Sanderson and Kupferberg, "Development and Evolution," 306.
29 *OED*, s.v. "proportion, n.," senses I.5 and I.6, revised June 2007.
30 Sanderson and Kupferberg, "Development and Evolution," 309.
31 Sanderson and Kupferberg, "Development and Evolution," 309.
32 Gervase Markham, *Country Contentments*, 4th ed., in *A way to get wealth* (1631, STC [2nd ed.] 17396), H3r/101. The pagination of this volume is not continuous.
33 John Donne, "Preached at Lincolns Inne," in *The Sermons of John Donne*, ed. Evelyn M. Simpson and George R. Potter (Berkeley and Los Angeles: University of California Press, 1957), vol. 3, sermon 3, 1, http://contentdm.lib.byu.edu/cdm/compoundobject/collection/JohnDonne/id/3153/rec/3. Text citations refer to page numbers from the digitized edition.
34 St. Justin Martyr, *The First and Second Apologies*, Ancient Christian Writers 56, trans. Leslie William Barnard (New York: Paulist Press, 1997), 36.
35 Justin, *The First and Second Apologies*, 36.
36 Athenagoras, *Embassy for the Christians, The Resurrection of the Dead*, Ancient Christian Writers 23, trans. Joseph Hugh Crehan (New York: Paulist Press, 1955), 84.
37 Athenagoras, *Embassy for the Christians, The Resurrection of the Dead*, 84, 90.
38 John Donne, "Preached White-hall, the first Friday in Lent," in *The Sermons of John Donne*, ed. Evelyn M. Simpson and George R. Potter (Berkeley and Los Angeles: University of California Press, 1959), vol. 4, sermon 13, 4, http://contentdm.lib.byu.edu/cdm/compoundobject/collection/JohnDonne/id/3181/rec/6.
39 Donne, "Preached at White-hall," 4.
40 Gessner devotes the fourth volume of his *Historiae animalium* to fish (1558); a fifth volume on snakes and scorpions was left incomplete and published posthumously (1587). Topsell includes no fish in his *Historie of Serpents* (1608, STC [2nd ed. 24124]).
41 William Shakespeare, *Hamlet*, ed. Ann Thompson and Neil Taylor (London: Arden Shakespeare, 2006).
42 *OED*, s.vv. "vellum, n.," "veal, n.1," and "velum, n.," accessed April 11, 2018.
43 *OED*, s.v. "fry, n.1;" *MED*, s.v. "frī(e, n.2)" http://quod.lib.umich.edu/cgi/m/mec/med-idx?type=id&id=MED17737, accessed April 11, 2018.
44 Taverner, *Certaine experiments*, 11.
45 Robbins cites Tilley S446 in *The Complete Poems of John Donne*, 443n250. (Morris Palmer Tilley, *A Dictionary of the Proverbs in England in the Sixteenth and Seventeenth Centuries* [Ann Arbor: University of Michigan Press, 1950]).

46 Robbins notes, "The muteness of fish was proverbial from ancient Greek times, whence the Pythagoreans, according to Plutarch, *Symposiaca*, regarded them as fellow disciples and abstained from eating them" (*The Complete Poems of John Donne*, 445n283). In entry F300, "As mute as a fish," Tilley cites an adage from Erasmus and several early modern examples (Tilley, *A Dictionary of the Proverbs*, 217).

47 "Erasmus in his letters often complains of ill health, of his aversion to fish, and of his stomach troubles" (Craig R. Thompson, *Collected Works of Erasmus: Colloquies*, vol. 1 [Toronto: University of Toronto Press, 1997], 102n45).

48 Desiderius Erasmus, *Collected Works of Erasmus: Colloquies*, vol. 1, 680. The colloquy is entitled "A Fish Diet." The two relevant adages are I i 2 ("*A piscibus abstineto* / Do not taste fish") and I v 29 ("*Magis mutis quam pisces* / As dumb as the fishes"), both in *Collected Works of Erasmus: Adages*, vol. 2, trans. Margaret Mann Phillips (Toronto: University of Toronto Press, 1982), 48–49 and 408–9.

49 Robbins, ed., *The Complete Poems of John Donne*, 87n57-8, 257-58n1-4.

50 Izaak Walton, *The Life of John Donne, Dr. in Divinity, and Late Dean of Saint Pauls Church London* (1658, Wing W668), 121. This description, which comes near the end of the "life," is prefaced with an account of Donne's monument—a story that, in its final revision, echoes the poet's observations about the subtle mechanisms of respiration. In the first printed version of the biography, the statue is admired by Dr. Fox, who contributed a hundred marks to its establishment and "lived to see as lively a representation of his dead friend as Marble can express" (in John Donne, *LXXX sermons preached by that learned and reverend divine, Iohn Donne, Dr in Divinity, late Deane of the cathedrall church of S. Pauls London* [1640, STC (2nd ed. 7038)], preface). For the 1658 revision, published on its own, Walton maintains those details and introduces a second particular observer: here, the statue is "indeed so like Dr. Donne, that (as his friend Sir Henry Wotton hath expressed himself) *it seems to breath[e] faintly, and Posterity shall look upon it as a kind of artificiall Miracle*" (120, original italics).

51 Bone and Moore, *Biology of Fishes*, 125.

52 Thomas Grubb, *The Mind of the Trout: A Cognitive Ecology for Biologists and Anglers* (Madison: University of Wisconsin Press, 2003), 36.

53 Grubb, *The Mind of the Trout*, 36.

54 Grubb, *The Mind of the Trout*, 37.

55 Grubb, *The Mind of the Trout*, 38.

56 *OED*, s.v. "smolt, v.2;" *MED*, s.v. "smolten, v." (http://quod.lib.umich.edu/cgi/m/mec/med-idx?type=id&id=MED41132), accessed April 11, 2018.

57 *OED*, s.v. "smolt, n.1," accessed April 11, 2018.

58 *OED*, s.vv. "seely, adj.," "silly, adj., n., and adv.," accessed April 11, 2018. More's usage comes from *The supplycacyon of soulys made by syr Thomas More knyght councellour to our souerayn lorde the Kynge and chauncellour of hys Duchy of Lancaster* (1529, STC [2nd ed.] 18092).

59 Leonard Mascall, *A Booke of fishing with Hooke & Line, and of all other instruments thereunto belonging* (1590, STC [2nd ed.] 17572), 27. Mascall's manual begins as a mere update of the *Treatyse of fysshyng with an Angle*, but the instructions for catching piscivorous animals (other than humans) are notable additions to the earlier text.

60 Mascall, *A Booke of fishing*, 32.
61 Mascall, *A Booke of fishing*, 32.
62 *OED*, s.v. "whale, n.," accessed April 11, 2018.

## References


Athenagoras. *Embassy for the Christians, The Resurrection of the Dead.* Ancient Christian Writers 23. Translated by Joseph Hugh Crehan. New York: Paulist Press, 1955.

Balcombe, Jonathan. *What a Fish Knows: The Inner Lives of Our Underwater Cousins*. New York: Scientific American / Farrar, Straus and Giroux, 2016.

Bone, Quentin, and Richard H. Moore. *Biology of Fishes*. 3rd ed. New York and Abingdon: Taylor & Francis Group, 2008.

Collins, Siobhán. *Bodies, Politics and Transformations: John Donne's Metempsychosis*. Farnham and Burlington: Ashgate, 2013.

Donne, John. *The Complete Poems of John Donne*, edited by Robin Robbins. Harlow, UK: Pearson Education, 2010.

———. "Preached at Lincolns Inne." In *The Sermons of John Donne*, edited by Evelyn M. Simpson and George R. Potter. Vol. 3, sermon 3. Berkeley and Los Angeles: University of California Press, 1957. http://contentdm.lib.byu.edu/cdm/compoundobject/collection/JohnDonne/id/3153/rec/3.

———. "Preached White-hall, the first Friday in Lent." In *The Sermons of John Donne*, edited by Evelyn M. Simpson and George R. Potter. Vol. 4, sermon 13. Berkeley and Los Angeles: University of California Press, 1959. http://contentdm.lib.byu.edu/cdm/compoundobject/collection/JohnDonne/id/3181/rec/6.

———. "Untitled [second sermon preached at the Hague]." In *The Sermons of John Donne*, edited by Evelyn M. Simpson and George R. Potter. Vol. 2, sermon 14. Berkeley and Los Angeles: University of California Press, 1955. http://contentdm.lib.byu.edu/cdm/compoundobject/collection/JohnDonne/id/3146/rec/2.

Englands Helicon. London, 1600. STC (2nd ed.) 3191.

Erasmus, Desiderius. *Collected Works of Erasmus: Colloquies*. Vol. 1. Edited by Craig R. Thompson. Toronto: University of Toronto Press, 1997.

——— *Collected Works of Erasmus: Adages*. Vol. 2. Translated by Margaret Mann Phillips. Toronto: University of Toronto Press, 1982.

Gessner, Konrad. *Historiae animalium, Liber IV, Qui est de Piscium & Aquatilium Animatium natura*. 1558. Frankfurt, 1604.

Grierson, Herbert J. C. "Donne's Love Poetry." In *John Donne: A Collection of Critical Essays*, edited by Helen Gardner, 23–35. Englewood Cliffs, NJ: Prentice Hall, 1962.

Grubb, Thomas C. *The Mind of the Trout: A Cognitive Ecology for Biologists and Anglers*. Madison: University of Wisconsin Press, 2003.

Hall, Brian K., and Marvalee H. Wake. Introduction to *The Origin and Evolution of Larval Forms*. San Diego, CA: Academic Press, 1999.

Hickman, Carole S. "Larvae in Invertebrate Development and Evolution." In *The Origin and Evolution of Larval Forms*, edited by Brian K. Hall and Marvalee H. Wake, 21–59. San Diego, CA: Academic Press, 1999.

Justin, Martyr, St. *The First and Second Apologies*. Ancient Christian Writers 56. Translated by Leslie William Barnard. New York: Paulist Press, 1997.

Maitland, Peter S. *The Hamlyn Guide to Freshwater Fishes of Britain and Europe*. London: Hamlyn, 1977.

Markham, Gervase. *Country Contentments*. 4th ed. In *A way to get wealth*. London, 1631. STC [2nd ed.] 17396.

Mascall, Leonard. *A Booke of fishing with Hooke & Line, and of all other instruments thereunto belonging*. London, 1590. STC (2nd ed.) 17572.

More, Thomas. *The supplycacyon of soulys made by syr Thomas More knyght councellour to our souerayn lorde the Kynge and chauncellour of hys Duchy of Lancaster*. London, 1529. STC (2nd ed.) 18092.

Rosch, Eleanor. "The Environment of Minds: Toward a Noetic and Hedonic Ecology." In *Cognitive Ecology*, edited by Morton P. Friedman and Edward C. Carterette, 3–27. San Diego, CA: Academic Press, 1996.

Sanderson, S. Laurie, and Sarah J. Kupferberg. "Development and Evolution of Aquatic Larval Feeding Mechanisms." In *The Origin and Evolution of Larval Forms*, edited by Brian K. Hall and Marvalee H. Wake, 301–77. San Diego, CA: Academic Press, 1999.

Shakespeare, William. *Hamlet*. Edited by Ann Thompson and Neil Taylor. London: Arden Shakespeare, 2006.

Taverner, John. *Certaine experiments concerning fish and fruite: practised by Iohn Tauerner Gentleman, and by him published for the benefit of others*. London, 1600. STC (2nd ed.) 23708.

Tilley, Morris Palmer. *A Dictionary of the Proverbs in England in the Sixteenth and Seventeenth Centuries*. Ann Arbor: University of Michigan Press, 1950.

Topsell, Edward. *The Historie of Serpents*. London, 1608. STC (2nd ed.) 24124.

*Treatyse of fysshynge wyth an Angle*. In *This present boke shewyth the manere of of hawkynge [and] huntynge and also of diuysynge of cote armours ...* [The Book of Saint Albans]. Westminster, 1496. STC (2nd ed.) 3309.

Walton, Izaak. *The Compleat Angler or, the Contemplative Man's Recreation*. 1st ed. London, 1653. Wing (2nd ed.) W661.

———. *The Life of John Donne, Dr. in Divinity, and Late Dean of Saint Pauls Church London*. London, 1658. Wing W668.

———. "The Life and Death of Dr Donne, Late Deane of St Pauls London." Preface to *LXXX sermons preached by that learned and reverend divine, Iohn Donne, Dr in Divinity, late Deane of the cathedrall church of St. Pauls London*, by John Donne. London, 1640. STC (2nd ed. 7038).

———. "The Life of Sir Henry Wotton." Preface to *Reliquiae Wottonianae*, by Henry Wotton. London, 1651. Wing [2nd ed.] W3648.

Wotton, Henry. "On a Banck as I sate a Fishing, A Description of the Spring." In *Reliquiae Wottonianae*, 524. London, 1651. Wing [2nd ed.] W3648.

# 3 The Fishing Lines of John Dennys

Printed posthumously in 1613, John Dennys's *Secrets of Angling* is the first known fishing poem in English literary history. Although angling manuals were being published long before the seventeenth century, Dennys's work is the first to set down fishing instructions in lines of verse, and it is thus the point of origin for a subgenre that flourishes even today. It would be hard to overstate the influence of *The Secrets of Angling*, and yet its author was not accurately identified for many generations.[1] Izaak Walton quoted the poem at length in all editions of his much more famous fishing treatise (starting in 1653), but never ascribed it to Dennys. Remarkably, in a centuries-long series of misattributions, the poem was once considered to be the work of John Donne.[2] The poem has been read consistently, in whole or in part, for the past four hundred years. Sections of Dennys's advice have regularly been quoted by fellow anglers—in pamphlets, books, magazines, and blogs—since the poem's first printing. The poem's major source is the medieval *Treatyse of fysshynge wyth an Angle*, which had been repeatedly edited and adapted by the time of Dennys's writing. It also borrows from William Samuel's *Arte of Angling*, a manual in dialogue form that would become Walton's primary inspiration for *The Compleat Angler*.[3] Dennys mingles material from the available literature on angling and boldly translates it into poetry; the generic transformation of piscatory knowledge from prose to verse is his greatest innovation.

This chapter examines the formal features of *The Secrets of Angling* and argues for a strong connection, advanced by the text itself, between sportfishing and poetic expression. When poet-anglers of the twenty-first century take for granted a certain sympathy between the very distinct recreational activities of writing verse and catching fish, they have generations of fellow writers standing behind them, similarly compelled to set their riverside observations down in measured lines. Marking the beginning of this tradition, *The Secrets* adopts elements of Ovid's *Metamorphoses* and Boccaccio's long verse narratives, but it draws the better part of its inspiration and information directly from the English countryside. Because it presents images from the banks of the River Avon, which runs through Dennys's home county of Gloucestershire, the poem partly belongs in the contemporary genre of river poetry. At the same time, by delivering detailed angling advice in verse, *The Secrets* sets itself apart from all existing categories. On the basis of its combined poetic and didactic value, Dennys and his publisher are keen to assert the originality of the treatise.

Aside from enshrining Dennys as the original English angler-poet, a careful reading of his work indicates a deepening of the early modern fascination with the cognitive faculties of fish. *The Secrets* moves well beyond the cursory references to piscine minds in the *Treatyse*, presenting readers with detailed commentary on the morphology and behavior of individual species. Here, angling is a way for human beings to relate to many kinds of fish; the sport is as varied as the species it pursues. Although evolutionary ecology wouldn't emerge until centuries after Dennys's death, his studies of fish in many ways correspond with that discipline's approach to animals and their environments. In particular, the field of cognitive ecology contributes ideas about how creatures look and learn; these findings let us take Dennys seriously in his account of how human and piscine minds interact at the river's edge. Cognitive ecologists identify four categories in which most animals must make decisions that contribute to their fitness: "feeding, predator avoidance, interactions with competitors, and sexual behavior."[4] In the angling scenes of *The Secrets*, fish alter their behavior in response to several factors, one of which is human interference with their habitats. Consistently placing the cognitive processes of fish under investigation, the text demonstrates that it is impossible to succeed as an angler without imagining the way fish think. At the same time, Dennys describes the systems of apprehension and analysis that allow anglers to catch particular species. Thus the poem looks into the minds of fish and humans alike. Although there is no immediate threat to our species' survival in the world of Dennys's recreation, the poem does concern itself with human adaptation, promising to turn its readers into fitter anglers and therefore better people. In teaching readers how to interpret and exploit the apparent preferences of fish, *The Secrets* becomes a study of learning on multiple levels. The poem presents detailed data about what fish see, like, and learn, and it asks readers to accept this knowledge in a new way.

## Angling in Verse

We ought to begin not in the stream but on the bank of Dennys's text, with the bold title and frontispiece intended to draw hopeful anglers into the poem. The woodcut depicts an elaborate riparian scene filled with life forms, geological phenomena, and visual information about the weather. A river runs through it, parallel to the printed text, as if to indicate that the treatise itself will carry the readers further upstream as they turn the

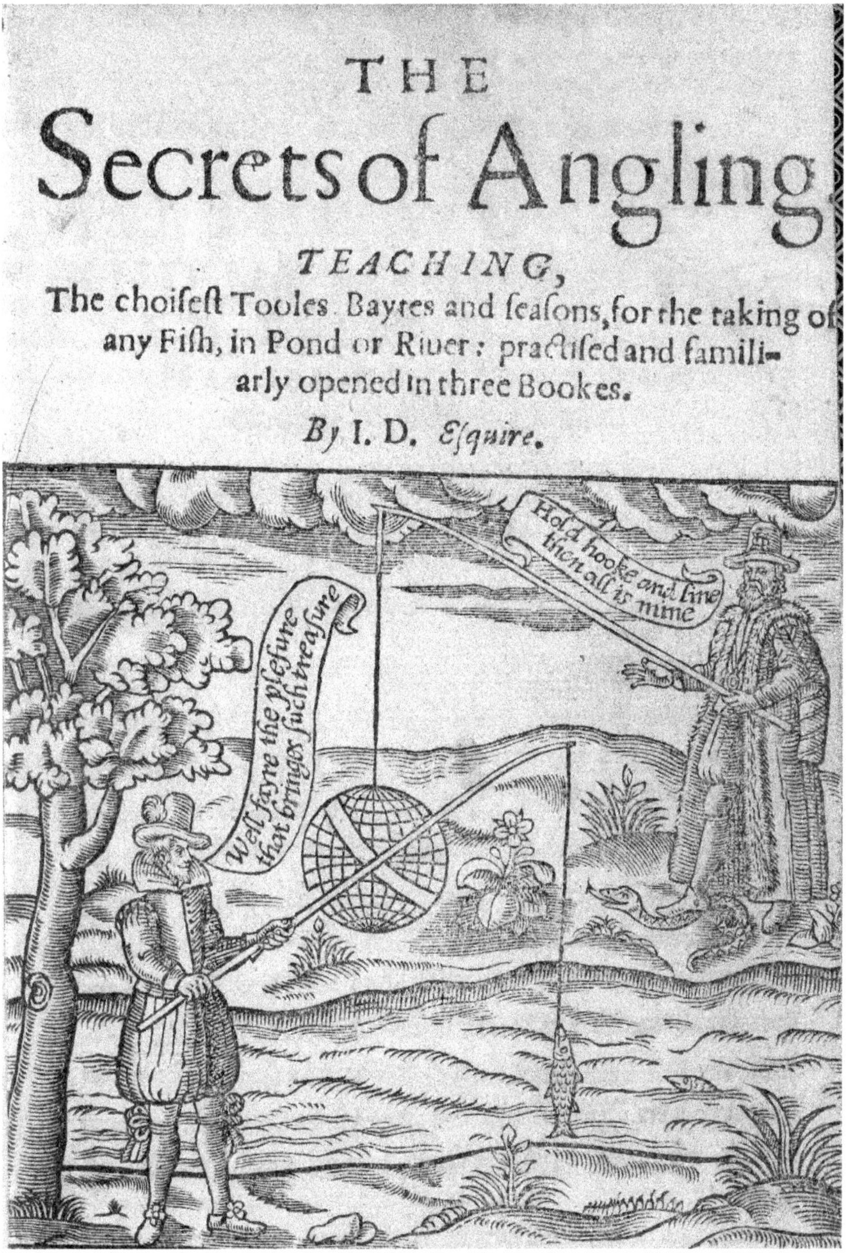

THE

Secrets of Angling

TEACHING,

The choisest Tooles Baytes and seasons, for the taking of any Fish, in Pond or River: practised and familiarly opened in three Bookes.

By I. D. Esquire.



*Figure 3.1* The title page of John Dennys's fishing treatise, 1613. Folger Shakespeare Library.

page. There are two human forms, quite plainly in competition with one another. They wear smug expressions and resplendent attire, and stand on opposite sides of the narrow stream with their rods extended toward the water from the left and right extremes of the frame. To balance the image even further, a banner of text hovers near the face of each gentleman, so they seem to speak—either aloud or in their minds—a pair of couplets. The verses themselves contend with each other, their respective banners tilting inward while their sentiments aim in different directions. The words of the man in the foreground (on this side of the river) are "Well fayre the plesure / that bringes such treasure," referring perhaps to the freshly caught fish at the end of his line. From the further angler, whose line holds only an empty, spherical trap, comes the rather more greedy-sounding "Hold hook and line / then all is mine." His trap looks like the globe, as if he very nearly does have the whole world in his hands, and yet it is bisected (from the reader's perspective) by the rod of the rival angler, whose simpler gear has proven more effective. Already, in the mild contest depicted on the title page, poetry and angling are interwoven.

The human beings are only part of the picture, however. There is an impressive diversity of plant life: tufts of grasses and burgeoning weeds, flowers extending their petals toward the sunlight, and a leafy tree. The evident undulation of the water corresponds with the sinuous line of the riverbank and the gentle rolling of the land in the foreground and the distance. The brightness of the clouds and the intensity of short shadows throughout the image suggest a midday setting. As one fish is being hauled up vertically, its neighbor comes swimming along, bearing a distinctly worried expression on its tiny face. Parallel to that visage but considerably more grinny is the face of a snake on the far bank, an animal that may lose its smile when it realizes that it is being trod upon by a human foot. The further angler and the reptile are cast in a performance of Genesis 3:15, where God tells the serpent, "I will put enmity between thee and the woman, and between thy seed and her seed; it shall bruise thy head, and thou shalt bruise his heel."[5] The illustration is packed with information attesting to an abundant ecosystem that features both vibrancy and violence.

Much in the image is distinct from what the poem offers: the couplets are not excerpted from Dennys's poem, there is no instruction in *The Secrets* on the building or deployment of spherical traps, and the speaker does not present his sport as an occasion for competition between gentlemen. Further, the instructor would certainly not approve of the showy garments bedecking either angler; they are too eye-catching to permit the catching of fish, especially on such a sunny day as this appears to be. In a section on "what garment is fittest," the speaker calls for clothes of "Russet … or gray, / Of colour darke, and hardest to discry; / That with the Raine or weather will away, / And least offend the fearefull Fishes eye" (B5v). But the woodcut nevertheless prepares the reader for an account of angling that is richly situated in an environment. The world of *The Secrets* teems with cultural and creaturely life. Most importantly, the treatise reveals the vibrant ecological ramifications of human language, including the rehearsal of received texts, the spontaneous composition of poetry, and the reading of angling instructions, all of which are on display in the reader's first encounter with the frontispiece.

Just as the woodcut assembles multiple cognitive systems into a single group portrait, the prefatory matter printed with the poem involves the minds of the poet, the publisher, the dedicatee, and the common reader in an effort to promote Dennys's project. The dedicatory epistle from the publisher to his friend "Mr. John Harborne" and the sonnet of praise from "Jo. Davies" are perfectly conventional, but they merit particular attention because of the way they represent the poem's worth to its readers. Both texts describe cognitive processes that parallel the mental work of the angler. "R. I.," the publisher, explains to his friend Harborne that he doesn't want a good poem, "sauouring of Art and Honesty," to go to waste; he is "loath to see a thing of such value lye hidden in obscuritie."[6] Although there is no explicit reference here to the lurking of fish under mud and silt, R. I. conjures a vividly physical image—the text is precious and wholesome, and should therefore be drawn out of its shadowy habitat and shared. The publisher also describes his search for an enthusiastic recipient of the poem: "I could not among my good friends, bethinke me of any one to whom I might more fitly dedicate it (as well for the nature of the subiect in which you delight as to expresse my love) than to yourselfe" (A3r). At the edge of a poem that is deeply invested in telling readers how to suit their bait to their prey, this contemplation of what might "delight" another person has special significance. Like bait offered to fish, published poetry reaches its target only if someone has bothered to imagine the desires that abide within some other (perhaps very different) mind.

The comparison between angling and literary production is much more explicit in Davies's sonnet, where the rhetorical effect of the poem turns the reader into a kind of catch. The "Angler," as Davies nicknames the poem, is itself a fisher that "takes, / With many a Line, well made, both Eares and Harts" (A4r). The important difference between the two recreations is that the angler's art is understood to make a fool of the fish, while the instructional poet "the skill-lesse skill-full makes." Dennys's poem will capture the interest of its readers, and they may be surprised at being drawn in by such "plaine Lines," but all will be to their benefit. In Davies's central pun, the *skill* of the poet—his reasoned, verbal defense of the sport[7]—becomes the skill of his readers. The etymology of *skill* suggests that practical knowledge grows out of division and discrimination; the old verb form of the word meant "to separate, divide."[8] As if to activate this obsolete signification, Davies presents an anatomizing image of the poem as a "Corpes" of information, "dissected" by the poet

(9). Incidentally, if the sonnet was composed by the prolific poet and writing-master John Davies, its emphasis on the persuasive and pedagogical value of *The Secrets* makes all the more sense. *That* Davies, as a teacher of handwriting, would have been particularly attuned to the marketability of specialized skills and the allure of lines well made. He taught penmanship because his poetry fetched little in the way of praise or pay.[9]

In both the prefatory sonnet and the dedicatory epistle, the poetic form of Dennys's instructions is defended. R. I. acknowledges that the subject of angling is here "reduced into the nature of a Poeme," but is certain that his dedicatee will "find it more briefly, pleasantly, & exactly performed, then any of this kinde heretofore" (A3v). Similarly, Davies declares, "Vpon an humble Subiect never lay / More proude, yet plainer Lines" (A4r). The concision of verse is thus connected to the practicality of Dennys's text. Rather than being useful in spite of its inventive form, *The Secrets of Angling* will, according to these writers, prove itself "both pleasant and profitable" through its versification (A3r). Its *skill* is thus properly understood in two concurrent ways, as both the detailed instruction it transmits and the artful division of that instruction into lines of verse. In fact, *skill* is a tidy metaphor for versification itself: a practice of discrimination by which verbal units are separated into measured lines.

But the compatibility of the poem's aesthetic qualities and its didactic function is not something these writers can take for granted. R. I. compares Dennys's text to popular poetry about husbandry, hunting, and hawking, while Davies whets the reader's appetite by presenting his praise in sonnet form. In these opening texts, as in Dennys's lines themselves, we are asked to receive *The Secrets* as something new: a work that argues for the didactic and mnemonic viability of poetic structure by delivering practical information, line by line, about the art of angling. R. I. and Jo. Davies insist, as many writers of prefatory matter might do, that the work they introduce constitutes a substantial body of new information. More aptly, though, they recognize technological advancement in the very form of the text. Dennys's poem, combining the long history of angling with the long history of poetry, is itself centrally concerned with the idea of technology. *The Secrets* is oriented toward the future of the sport. On the other hand, when it comes into print in 1613, it emerges into a verbal landscape that is already rich in representations of angling. In one way or another, Dennys has to deal with the existing network of figures that have exploited the metaphorical potential of his sport in everyday speech and thought.

Connecting the human body to an unknown liquid environment, the angler's line can be read as a convenient metaphor for the cognitive operations upon which the sport relies. Virginia Woolf uses it this way in *A Room of One's Own*:

> Thought—to call it by a prouder name than it deserved—had let its line down into the stream. It swayed, minute after minute, hither and thither among the reflections and the weeds, letting the water lift it and sink it, until—you know the little tug—the sudden conglomeration of an idea at the end of one's line: and then the cautious hauling of it in, and the careful laying of it out?[10]

For Woolf and for us, angling is a powerful and familiar source of meaning. As we have seen in Davies's sonnet, where lines can "take" our ears and hearts, metaphors that compare mental labor to the efforts of an angler are at least as old as the genre of the fishing poem. Like Woolf, the speaker of *The Secrets* will follow a conceptual line as it moves beneath the surface of the water. This line remains much more literal in Dennys's text, but it is nevertheless a device for looking deeper and a way to gather ideas. At the same time, we might remember Davies's comparison of the fishing line to the poetic line, and recognize in the formal challenges of Dennys's task a further demand upon the imaginations of writer and reader alike.

Although Dennys does not use the fishing line as a metaphor, his instructions are situated in a cultural context where figurative uses of angling are common and varied. By the time *The Secrets* was being composed, its subject was already widely deployed in non-literal ways that would necessarily contribute to a cultural understanding of the sport. If we accept a basic premise of conceptual metaphor theory, that "our conceptual system is shaped by our constant successful functioning in the physical and cultural environment,"[11] we can read the literal instructions of *The Secrets* in relation to both a physical environment that separates human and piscine life, and a cultural environment filled with non-literal representations of the sport. Angling metaphors seem to gather their force from the strong distinction between human and piscine bodies; in that sense, they belong in George Lakoff and Mark Johnson's category of "orientational metaphors," which are based on "spatial orientations [that] arise from the fact that we have bodies of the sort we have and that they function as they do in our physical environment."[12] For example, the idea of "fishing for compliments" would make little sense if angling did not entail the casting of a line from the land into someone else's habitat. That fish prefer not to be caught, that human beings can catch them without swimming, and that we eat them afterwards—all of these facts contribute to the emergence and resonance of angling metaphors.

But while concepts that come from our bodily experience of the physical world are, according to Lakoff and Johnson, "concepts that we live by in the most fundamental way," they are by no means independent from cultural meaning: "What we call 'direct physical experience' is never merely a matter of having a body of a certain sort; rather, *every* experience takes place within a vast background of cultural presuppositions."[13] It is probably impossible, then, to distinguish between physical and cultural environments. Instead, we can remember Lawrence Buell's invocation of "a myth of mutual constructionism" in which the physical environment and human culture are constantly shaping and reshaping one another.[14] As an aspect of cognition that is influenced by the way our bodies move through the world, metaphor is not only an airy phenomenon hovering somewhere above the landscape, but a tangible part of the environment in which we live. Non-literal representations of human activities like angling participate in our understanding of ourselves, our habitats, and our relationships with other species. They form part of the "vast background" behind a practical text like Dennys's angling poem. Ways of thinking (both literal and metaphorical) may be difficult to chronicle, but they are nevertheless ecological realities.

The principles of cognitive ecology, with its focus on how information travels between individuals, can be used to argue for the material force of poetry, especially if that poetry aims to orchestrate an interaction between species. In an evolutionary study of signal form, describing how plants and animals communicate through color patterns, songs, and facial expressions, Magnus Enquist and Anthony Arak emphasize the importance of nearby minds: "The receiver's brain is as much a part of the signal's environment as the external medium through which the signal must first travel to reach the receiver's sensory organs."[15] The spots on the trout are related to the water and pebbles of the stream in which the animal is camouflaged, but they are also involved in transactions with the eyes and brains of other animals: the human, the osprey, the fellow trout. When Gerard Manley Hopkins notices "rose-moles all in stipple upon trout that swim" and sets that observation down in verse, he adds something to the environment in which both species live.[16] When John Dennys writes his angling treatise, he too contributes to an ecosystem in which human thoughts tend toward both the admiration and the capture of fish.

## Vernacular Technology

*The Secrets* is nothing if not literary. Dennys neatly divides the text into three books, and in the course of a hundred and fifty-one stanzas of ottava rima, he moves back and forth from the general to the specific, addressing everything from the moral benefits of angling to the criteria for choosing a rod-worthy branch of hazel wood. From its very first lines, *The Secrets* dallies with the epic mode and the tenor of early modern sermonizing:

> Of Angling, and the Art thereof I sing,[17]
> What kinde of Tooles it doth behoue to haue;
> And with what pleasing bayt a man may bring
> The Fish to bite within the watry waue.

A work of thankes to such as in a thing
Of harmlesse pleasure, haue regard to saue
Their dearest soules from sinne; and may intend
Of pretious time, some part thereon to spend.
(B1r)

The poem seems to open with an invocation, but the matter of supernatural inspiration is delayed to allow for a dedicatory gesture: the text is given in exchange for the generous bestowal of time that readers will make on behalf of their souls. In ambiguously signaling a "thing / Of harmlesse pleasure," the speaker refers to both the practice of angling and the poem that describes that practice, two sites of activity juxtaposed in a broader category of virtuous recreation. To read about angling is already to engage in the "art," and to make one's way toward salvation.

In the double invocation that follows, Dennys draws on literary and cultural traditions that envision vibrant ecological diversity. Beginning with Ovidian nymphs, meadows, and birds, the invocation moves swiftly to the real-life rivers of England and the finny creatures who people them:

You Nymphs that in the Springs and Waters sweet,
Your dwelling haue, of euery Hill and Dale,
And oft amidst the Meadowes greene doe meet
To sport and play, and heare the Nightingale;
And in the Riuers fresh doe wash your feet,
While Prognes sister tels her wofull tale:
Such ayde and power unto my uerses lend,
As may suffice this little work to end.

And thou sweet Boyd that with thy watry sway,
Dost wash the cliffes of Deington and of Weeke;
And through their Rockes with crooked winding way,
Thy mother Auon runnest soft to seeke:
In whose fayre streames the speckled Trout doth play,
The Roche, the Dace, the Gudgin, and the Bleeke.
Teach me the skill with slender Line and Hooke
To take each Fish of River, Pond, and Brooke.
(B1r-v)

By starting with a classical source and shifting toward home, Dennys is able to remind his readers that England has its own mythology of rivers. This Englishing tendency is evident even before we approach the references to the Boyd and the Avon. Instead of opting to place "Philomel" at the end of the fourth line, where it would provide a slant rhyme for the foregoing "Dale" and a better match for the "Progne" named two lines later, Dennys lets an English bird perch in the Greek story. The fleeting allusion is then placed at a further remove, with the "wofull tale" of Philomela's suffering heard only as a distant song while the nymphs go about their business. These figures, too, are soon dismissed.

When the focus shifts from bodies of nymphs to bodies of water, the Greek myth is traded out for something more local and immediate. The tragic narrative of Procne and Philomela fully recedes, and another family unit captures the speaker's attention: the Boyd, a brook in Dennys's home parish of Pucklechurch, and its "mother," the Bristol Avon.[18] While the nymphs in the first half of the invocation may be able to offer "ayde and power," it is the rivers themselves that will dictate the practical information the speaker needs. Unlike the waters of the opening stanza, the Boyd and the Avon are not mere footbaths, but powerful forces with their own native populations. Nymphs may dip their feet wherever they choose, but rivers are real agents of geological change; they "wash … cliffs" and hold "watry sway." They are also teeming with life, and are taken to be authorities on the species they hold. The speaker only briefly appeals to the nymphs for their help before turning to the Boyd with a renewed request: "Teach me."

The second half of the invocation sounds like a revision, an attempt to get much closer to English fish by entering their habitats. With the naming of five distinct species in the waters of the Boyd, the speaker readies his audience for the more detailed piscine portraits that will appear later in the poem. *The Secrets*, like any useful angling manual, takes its readers underwater so they can begin to identify each individual species by their shape and behavior. This movement through what Dennys calls "the watry brimme" (D1r) is repeated again and again as the poem delivers explicit instructions on how to catch various kinds of fish. Getting from land ("Hill and Dale") to water ("fayre streames") is essential, but it's impossible without the development of certain habits of mind. Above all else, Dennys's advice insists that the angler must cultivate faculties of observation and imagination that can overcome the undeniable elemental difference between our world and the world of fishes. So when the nymph's foot dips into the stream, beyond the brim, Dennys's speaker follows and forgets it.

More important than the inheritance of classical traditions is another *line*: the storied life of nearby rivers. Without abandoning Ovid altogether, the speaker embarks on a documentary and didactic project that has its most essential sources in the water. While Dennys plainly borrows his verse form from Boccaccio and various snippets from Ovid, he also enters a conversation with native literature by writing poetry that is deeply invested in the representation of English rivers. In places, *The Secrets* builds on the tradition of Tudor river poetry—works like John Leland's "Cygnea Cantio" and William Vallans's "A Tale of Two Swannes," topographical poems which use the central figure of a personified river as a way to describe a particular region and its people.[19] In a related category of its own is Michael Drayton's enormous *Poly-Olbion*, an ambitious project begun in the last years of the sixteenth century and printed in two parts (1612 and 1622). Dennys was dead by the time the first part came into print, but it could be that the publisher of *The Secrets* was encouraged by the grand fluvial vision of Drayton's poem, which maps out England and Wales through lively animations of their rivers. Although it features no scene of angling, the first part of *Poly-Olbion* does praise several rivers for how "fishfull" they are, and it includes a wonderful description of salmon migrating up the River Tivy.[20] Keeping such works in mind, we can recognize in Dennys's text a reflection of the broad cultural concern, apparent in many contemporary genres, with the nature of rivers. Early modern England understood itself as a riparian nation, a place that had been shaped by fluvial phenomena and would continue to exploit the potential of its waterways. Andrew McRae explains, "For many writers, especially in the early decades of the seventeenth century, rivers offered at once a way of organizing spatial description and a vehicle for the generation of narrative and mythology."[21]

In a long view of the river as a topos through which ancient and recent cultures have represented the faculty of human thought, early modern riparian poetry is, for Wyman Herendeen, not so much an innovation as a continuation:

> In fact, the topographical poem of the seventeenth century is very much the product of this [classical and Judeo-Christian] tradition of using the rivers to explore the physical world for our moral and intellectual, as well as physical orientation. The rivers continue to serve as borders joining the realm of the mind and language, and that of the senses, expanding the frontiers of understanding.[22]

The inaugural work in a new subset of riparian writing, *The Secrets* inherits the whole metaphorical potential of rivers. At the end of each book, the speaker returns to an image of teacher and student making a journey by boat; the text itself is a stream of information. If the river somehow allows for a kind of "orientation," helping a human mind understand itself, then the art of angling figured forth in Dennys's poem is a quintessential riverside activity. The intellectual and the physical may become intertwined in other outdoor meditations, but in angling literature, which is always a contemplation of bodies and minds, they are necessarily indivisible.

In McRae's reading of riparian writing, a "powerful fantasy of circulation" emerges;[23] rivers represent mobility and are thus instrumental in various discourses of social change. For Maggie Kilgour, river poetry promotes a vision of Britain as "an independent whole that yet within itself was able to contain diversity."[24] By comparison with these well-founded accounts of how early modern writers often deployed images of England's waterways, a reading of Dennys seems to offer up a *de*politicized river. *The Secrets* is certainly not polemical, but in its rapid and repeated shifts from Mediterranean authorities to homegrown knowledge, it argues for the inherent value of English rivers as sources of information just waiting to be sounded. Before sharing his knowledge of riparian species, the speaker devotes six stanzas to a detailed roster of what he *won't* be describing. The list is addressed to a powerful but irrelevant god:

> Here, O Neptune, that with triple Mace
> Dost rule the raging of the Ocean wide;
> I meddle not with thy deformed race
> Of monsters huge, that in those waues abide.
> (C3v)

Dennys's sport is characterized by its distance from the ocean, its preference for "the fresh," and its small, specialized equipment ("For well I thinke my Hookes would not suffise, / Nor slender Lines, the least of these to holde" [C4v]). Again, the language moves gradually away from the classical, beginning with an apostrophe to Neptune but ending with a catalog of English fish: "The little Roach, the Menise biting fast, / The slymie Tench, the slender Smelt and round, / ... And many more that in fresh waters are" (C4v). Here, angling is a vernacular sport, suited to the geography of England and satisfied with its humble yield. It is significant that the instructional parts of Dennys's text owe nothing to continental sources. Like the species he portrays, the poet gathers his information from the environment in which he lives. Thus he contributes to "the image of the rivers of a Britain surrounded by a sea," which, in Kilgour's terms, "suggested a national independence and wholeness."[25]

Beyond this subtle evocation of nationhood, the text offers images of communal living that feature both mobility and diversity. When a fly is dropped on the surface of the water, "the Chub, the Roche, the Dace, and Trout, / To catch thereat doe gaze and swimme about" (C7r). Although Dennys's fish are not political, they are seen to move happily through a shared habitat, and their behavior is everywhere compared to the actions of human beings. It's no real stretch to recognize in the swift and deliberate motions of these creatures an extension of the "fantasy of circulation" which McRae ascribes to contemporary descriptions of rivers. Read one way, fish are little synecdoches of their environments; they have courses, secrets, and caprices, just as rivers do. To treat Dennys's new genre of angling poetry as an offshoot of Tudor river poetry is to recognize that English rivers have both surface and depth—they can carry goods, and they often carry special significance for the human populations living on their banks, but they also carry life itself.

At the end of the first book of *The Secrets* is a section bearing the ambiguous title "The Author of Angling," a phrase that could refer either to Dennys himself as the writer of fishing instructions, or to some other figure who is understood to be the originator of the sport. Here the speaker turns explicitly to Ovid and complains that not all the information he seeks can be found in that venerable source. He doesn't know who exactly invented angling or when, because "Ovid wrote not all that story out" (B8r). In order to provide an origin myth that will satisfy "the willing Readers eare," the speaker embellishes the story of the flood from the *Metamorphoses*. After a new crop of people has sprung up to repopulate the postdiluvian world, one man finds a way to fend off hunger, and he shares his discovery:

> Then did Deucalion first inuent
> The art of Angling, and his people taught the same,
> And to the Woods and groues with them he went
> Fit tooles to finde for this most needfull game;
> There from the trees the longest ryndes they rent,
> Wherewith strong Lines they roughly twist and frame,
> And of each crooke of hardest Bush and Brake,
> They made them Hookes the hungry Fish to take.
> (C1v)

Aside from the threat of starvation, none of this comes from Ovid. The introduction of angling into the world of the *Metamorphoses* is Dennys's invention, as is the emphasis on teaching and the assertion that fish were much easier to catch back then: "The fish as yet had felt but little smart, / And were to bite more eager, apt, and bold" (C2r). Thus Deucalion and Dennys are folded together to become "the author of angling;" without Dennys's rewriting of Ovid, Deucalion would have no place in this piscatory poem, and without Deucalion, Dennys would have no point of origin, real or imagined, for the subject of his treatise. By inventing an inventor, the poet can participate in the whole history of the sport.

In *The Secrets*' account of the familiar myth of decline, fishing technology runs against the current: "So worse and worse two ages more did pass / Yet still this Art more perfect daily grew" (C2r). As human beings gain knowledge and power, fish become increasingly wary and difficult to catch. Modern biology would call this an example of coevolution: two distinct species adapt as a result of their interaction. On the human side, technology has to advance to meet the needs of each successive age of anglers. The first hooks are made of botanical material; then come silver and brass, then "hardest steele." Lines, once mere strips of bark, are ultimately made of "Silke and subtile hayre," and rods become increasingly slender, supple, and light. The speaker describes state-of-the-art equipment, and imagines no further improvement: "And so this Art did in the end attaine / Vnto that state where now it doth remaine" (C2v).

Despite the apparent perfection of the tools, the sport can yet be improved on another front. In the context of a didactic text, one that brandishes the word "TEACHING" across its title page, it is significant that Dennys makes Deucalion a *teacher* of angling. Since each successive age of humanity had not only new tools but also new skills to master in the pursuit of fish, the transmission of knowledge is necessarily one of the piscatory innovations added to the postdiluvian story. Dennys presents his poem as both an inheritor of ancient wisdom and a pioneering set of instructions. Becoming an "author of angling" in his own way, Dennys offers future generations of sportfishers something that has never been seen before: a collection of information in the fictional tradition of Deucalion, set down in lines for the very first time. The impetus behind this innovation, according to the logic of the revised myth of decline, is the steadily increasing cleverness of fish, without which there would be no need for an elaborate treatise in the first place. The hunted species, in all their wily evasion of human hooks, have paradoxically encouraged the emergence of a text that will endanger them. As *The Secrets* moves on from this origin story, it continues to show that an angler's success depends upon the careful analysis of how piscine minds function within their home environments.

## Looking and Learning

Taking its lead from both the spirited ecology of the *Treatyse* and the conventions of Tudor river poetry, *The Secrets* represents angling as a small part of a much larger ecosystem, focusing at times on the details of the sport but often zooming out, like the woodcut on the title page, to situate the river in a vast riparian zone that includes a wide range of plant and animal life. To refute the argument of "some youthful Gallant" that fishing is fruitless and stupid (B6r), the speaker launches into a defense in which he claims for angling a tremendous moral value, contrasted with the deleterious pursuits of cardplayers and womanizers.[26] The strong distinction between indoor and outdoor sports is here emphasized through an elaborate description of what the angler observes on a typical fishing excursion. He begins and ends with fish, but the space between is filled with reflections on life forms, topography, and weather. While those poor sots are stuck inside with cards, drink, and smoke, our angler enjoys freedom of movement:

> Let them that list these pastimes then pursue,
> And on their pleasing fancies feed their fill;
> So I the Fields and Meadowes greene may view,
> And by the Riuers fresh may walke at will …
>
> (B7r)

He lists flowers, naming their hues and echoing this image with a description of the "sundry kindes of painted collours" that appear at sunrise. A whole landscape is encompassed in his account of "hills and Mountaines … / plaines … / riuers …, / The surging sea beneath the valleys low, / The valleys sweet, and lakes that louely flow." There are woods filled with birdsong, and meadows in which flowers are "intermixt" with the grass like fish in a brook (B7v).

The speaker reaches out to include all of these things in the imagined purview of an angler who stands "on the pleasant Brinke" of whatever river (B7r), so that angling itself becomes the center of a practice that is not only observational but devotional. Gazing at the "strange and wonderfull" aspects of his surroundings, the fisher "Fram[es] thereof an inward contemplation, / To set his thoughts from other fancies free" (B7v). In his admiration for earthly phenomena, he is paradoxically compelled to think on higher things: "Whiles he looks on these with joyfull eye, / His minde is rapt above the starry skye" (B7v). In prizing the sport's ability to draw the angler into a contemplation of his environments, both physical and spiritual, *The Secrets* echoes its sources and further entrenches a claim that will remain a focus of angling literature throughout the early modern period (right through Izaak Walton's treatise on "the contemplative man's recreation") and beyond.

Although Dennys's speaker seems to refer only to a metaphysical upward gaze, in which the mind but not the eye is "rapt," he also talks about a very literal staring straight up: he looks at "the goodly compasse of the loftie Skye" (B7r), in an orientation that is opposite to the expected pose of the angler. But looking up and looking down appear to be equally viable gestures of both practical observation and religious devotion; the fisher can "see [his] Quill and Corke downe sinke, / With eager bit of Barbill, Bleike, or Dace: / And on the World and his Creator thinke" (B7r). His thoughts fly up while his eyes remain below. Dennys's vigorous defense of the sport asks us to imagine the angler quietly taking in his surroundings, casting his eyes up, down, and all around, and keeping in mind the higher power that has framed the scene of his recreation.

More than anything else, *The Secrets* delineates a program of visual work, a system of looking and watching that works to transcend the boundary between land and water. The angler stares into the river to see what fishes see, and manipulates visual stimuli in order to elicit certain responses. This fundamental aspect of the sport—its dependence upon the observation of how other animals visually apprehend their environment—is emphasized in the second book of *The Secrets*, where Dennys's speaker provides species-by-species instructions on how to catch particular kinds of fish. Here, each species is the subject, with its respective angler, of a discrete tableau. An imagined addressee, the nymph Thalia, is invited to watch several scenes unfolding at once. As the speaker prepares to deliver multiple lessons on looking, he uses a deictic mode to conjure concurrent moving images and direct our imaginative gaze:

> Sit then Thalia on some pleasant banke,
> Among so many as fair Auon hath,
> And marke the Anglers how they march in ranke,

Some out of Bristoll, some from healthfull Bath;
How all the Riuers sides along they flanke,
And through the Meadowes make their wonted path:
See how their wit and cunning they apply,
To catch the Fish that in the waters lye.
(C5r)

Implicitly, the final lines of this stanza compare the fishes' habitat to the mind of the angler. From the imaginary riverbank, which is here analogous to the reader's position at the edge of the text, Thalia can see neither the fish in the water nor the "wit and cunning" in the angler's intellect. The sections that follow will bring the invisible into view.

Each of the fishing scenes painted by Dennys is entitled as though it were a dedication to the species in question: "For the Goodgion," ... "For the Roche," and so on. In much the same way that the publisher asserts, in the dedicatory epistle, that he has thought about his friend's interest in angling ("the subject in which you delight"), the detailed instructions in the second book are organized around the preferences of each species. The angler who wishes to catch pike will need live bait, "For as for baites that dead and dull doe lye, / They least esteeme and set but little by" (C8r). Sewant and flounder "shunne the swifter streame" and lurk in the depths of the river (C7r). A fisher's success depends upon the consideration (and, to some extent, the memorization) of habits specific to each kind of fish, a pattern of recognition that takes piscine behavior to be motivated by the same sort of discernment and volition we expect from fellow humans. The fish "esteeme" certain things and "shunne" others, much as we do.

In sound pedagogical fashion, the speaker begins with the species who are most readily caught and gradually advances to the largest and most difficult fish, describing in each case an angler who is expert in securing his particular catch. A deictic gesture opens each successive scene: "Loe, in a little Boate where one doth stand;" ... "Then looke where as that Poplar gray doth grow;" ... "See where another hides himselfe as slye" (C5r, C6r, C7r). The reader is asked to picture the position of both the angler and his prey. But that's not all—these instructional images gaze into the minds of both animals, and give equal time to their respective cognitive operations. Although the speaker's hopes fall clearly on the side of the hunter, his attention is trained in equal measure on the imagined thoughts of the hunted. His relentless anthropomorphization of the fish may limit his ability to reckon with the particularities of their thinking, but he's not wrong to compare human and piscine cognition, or to treat neuronal processes as a dynamic feature of the environment in which any animal lives.

All of the animals depicted in the vignettes of the second book are engaged in learning. The anglers and the fish are equally finding their way toward a food source, while the reader (represented by Thalia) is gathering essential information from a distance. In cognitive ecology, learning is a form of "phenotypic plasticity:" an observable, reversible change in an animal's behavior as a result of its experience.[27] Far from representing knowledge acquisition as the special province of rational human beings, *The Secrets* lays out a *world* of learning. In summertime, the angler should let his fly fall repeatedly on the surface of the water so as to entice the fish: "With quicker hand, and with more ready skill / Let fall your flye, and now and then remoue, / Which soone the Fish will finde and better loue" (D2v). Here, what appeals to the fish is not only the tempting fly but also the feeling of having caught something. The prey, like its predator, has developed "more ready skill" in pursuit of its chosen food. This image of minds interacting in an environment corresponds with recent claims about how cognitive processes evolve. Like the cognitive ecologists who study honeybees and fruit flies, Dennys compares the learning of so-called "simple" creatures to that of big-headed humans, and finds that we are not as exceptional as we might have thought. In the context of *The Secrets*, the similarities between human and piscine cognition make the sport of angling possible, and the recreation itself steadily contributes to a store of communally held human understanding.

The third book of Dennys's poem is mostly taken up with a catalog of "The qualities of an Angler," an elaboration of a list of virtues from William Samuel's 1577 treatise *The Arte of Angling*.[28] In *The Secrets*, each essential characteristic is given its own stanza. Following Samuel's sequence, the last of these is aptly devoted to "memory," enjoining the hopeful angler to bear in mind "each needfull thing that he must occupy, / And not to stand of any want in doubt, / Or leaue something behind forgetfully" (D7r). As if to challenge the reader to exercise this very faculty, the text then wanders in a different direction, providing several pages of advice on the best and worst times to angle, along with a list of species' preferred habitats. Finally, near the end of the poem, the speaker reasserts the central importance of that final virtue by delivering a stanza that the reader is meant to commit to memory: a rhymed list of all the equipment that is needed for a successful day of angling. The lines are explicitly introduced as a mnemonic device ("Here shall he haue to help his memory, / A lesson short of every wants supply") and they neatly summarize the more detailed roster of necessary gear that is provided at the beginning of the treatise. Packed with assonance, alliteration, and parallelism, this penultimate stanza fairly represents the orderliness and artifice of Dennys's whole project:

> Light Rod to strike, long line to reach withall,
> Strong hooke to holde the fish he haps to hit,
> Spare Lines and Hookes, what ever chance doe fall,
> Baites quicke and dead to bring them to the bit,
> Fine Lead and Quils with Corks both great and small,

Knife, File and thred, and little Basket fit,
Plummets to sound the depth of clay and sand,
With Pole and Net to bring them safe to land.
(E2v)

The formal features that make this sequence literary are the very qualities that contribute to its effectiveness in the field. Rhyme, meter, and diction make the text memorable, so the verse becomes a tool that can be brought from home to the river. In addressing the particular problem of having to keep track of so many distinct pieces of gear, and supplying a solution to that problem in the same poetic form used throughout the treatise, Dennys draws the reader's attention once again to the didactic function of the entire work. Not only this summative stanza but rather the whole sweep of *The Secrets* is given over to the relation of practical information. Dennys's orientation toward an imaginative engagement with rivers and fish, his treatment of learning as a feature of embodied existence, and his technical virtuosity show that there is no meaningful distinction to be made between the practical and the literary, the substance and the form, the angler and the poet.

## Notes

1 Nicholas D. Smith, "Dennys, John (*d.* 1609), poet" *DNB* (September 2004), https://doi.org/10.1093/ref:odnb/7510.
2 This attribution comes from Robert Howlett, in *The Angler's Sure Guide* (1706). (Smith, "Dennys, John.")
3 [William Samuel], *The Arte of Angling* (1577), ed. Gerald Eades Bentley (Princeton, NJ: Princeton University Press, 1956).
4 Reuven Dukas and John M. Ratcliffe, eds., introduction to *Cognitive Ecology II* (Chicago: University of Chicago Press, 2009), 3.
5 Robert Carroll and Stephen Prickett, eds., "The First Book of Moses, Called Genesis," in *The Bible: Authorized King James Version*, Oxford Biblical Studies Online, accessed April 16, 2018, http://oxfordbiblicalstudies.com/article/book/obso-9780192835253/obso-9780192835253-div1-13.
6 R. I., dedicatory epistle prefacing I. D. [John Dennys], *The Secrets of Angling* (1613, STC [2nd ed.] 6611), A3r. Text citations of *The Secrets* and its front and back matter refer to signature numbers in this, the first edition of the poem.
7 *OED*, s.v. "skill, n.1," accessed April 16, 2018.
8 *OED*, s.v. "skill, v.1," accessed April 16, 2018.
9 P. J. Finkelpearl, "Davies, John (1564/5–1618), poet and writing-master," *DNB* (September 2004), https://doi.org/10.1093/ref:odnb/7244.
10 Virginia Woolf, *A Room of One's Own* (New York: Harcourt, 1929 [reprinted 1989]), 5.
11 George Lakoff and Mark Johnson, *Metaphors We Live By* (Chicago: University of Chicago Press, 1980), xii.

12 Lakoff and Johnson, *Metaphors We Live By*, 14.
13 Lakoff and Johnson, *Metaphors We Live By*, 57.
14 Buell, *Writing for an Endangered World*, 6.
15 Magnus Enquist and Anthony Arak, "Neural Representation and the Evolution of Signal Form," in *Cognitive Ecology*, ed. Reuven Dukas (Chicago: University of Chicago Press, 1998), 24.
16 Gerard Manley Hopkins, "Pied Beauty," in *Poems and Prose*, ed. W. H. Gardner (1953; repr., London: Penguin, 1985), 30. In the summer of 1864, a young Hopkins sketched an image of himself in his diary, giving it the caption "Gerard Hopkins, reflected in a lake." The drawing's perspective appears to be opposite to that of "Pied Beauty"—it is as if the viewer, like a trout, is glancing up at the poet from the water below. In *The Collected Works of Gerard Manley Hopkins,* vol. 3: Diaries, Journals, and Notebooks, ed. Lesley Higgins (Oxford: Oxford University Press, 2015), 211.
17 John McDonald finds here a humorous echo of the first line of the *Aeneid*: "Of arms and the man I sing" in *The Origins of Angling* (New York: Doubleday, 1957), 22.
18 Smith, "Dennys, John (*d.* 1609)." There are four other rivers that go by the name of Avon in England alone; Dennys's is often referred to as the "Bristol Avon" or the "Lower Avon" to distinguish it from these others, which include, of course, Shakespeare's Avon in Warwickshire. Samuel Lewis, *The Book of English Rivers* (London: Longman, 1855), 17–31.
19 Andrew McRae, "Fluvial Nation: Rivers, Mobility and Poetry in Early Modern England," *English Literary Renaissance* 38 (2008): 507.
20 Michael Drayton, *Poly-Olbion* (1612, STC [2nd ed.] 7226), 27, 83, 88. The second part (1622, STC [2nd ed.] 7229) finds the Trent delivering an extended brag about its ample and varied stores of fish (120–21).
21 McRae, "Fluvial Nation," 507.
22 W. H. Herendeen, "The Rhetoric of Rivers: The River and the Pursuit of Knowledge," in *Studies in Philology* 78, no. 2 (Spring 1981): 125.
23 McRae, "Fluvial Nation," 508.
24 Maggie Kilgour, "Writing on Water," *English Literary Renaissance* 29 (1999): 285.
25 Kilgour, "Writing on Water," 285.
26 The strong distinction Dennys makes between sportfishing and recreational sex is echoed in many places. A recent example is Paul Greenberg's book *Four Fish: The Future of the Last Wild Food*, where the author recalls his own adolescence: "The desire to pursue fish and the desire to pursue females of your own species are inversely proportional. The fishing jones waxes from about age seven until sixteen or so and then abruptly withers in the harsh hormonal light of adolescence.... The prime directive of life has shifted, and dusk no longer conjures the possibility of seabirds diving into a school of breaking game fish, but rather the moment when perfume and perspiration waft into the air, intermingled" (New York: Penguin, 2010), 6.
27 Dukas, "Evolutionary Ecology of Learning," *Cognitive Ecology* (1998), 129.
28 William Samuel, *The Arte of Angling*, 32–33.

## References

Buell, Lawrence. *Writing for an Endangered World: Literature, Culture, and Environment in the U.S. and Beyond*. Cambridge, MA: Harvard University Press, 2001.

Carroll, Robert and Stephen Prickett, eds. "The First Book of Moses, Called Genesis." In *The Bible: Authorized King James Version*. Oxford Biblical Studies Online. www.oxfordbiblicalstudies.com/article/book/obso-9780192835253/obso-9780192835253-div1-13.

[Dennys, John]. *The Secrets of Angling*. London, 1613. STC (2nd ed.) 6611.

Drayton, Michael. *Poly-Olbion*. London, 1612. STC (2nd ed.) 7226.

———. *The second part, or a continuance of Poly-Olbion from the eighteenth song*. London, 1622. STC (2nd ed.) 7229.

Dukas, Reuven. "Evolutionary Ecology of Learning." In *Cognitive Ecology: The Evolutionary Ecology of Information Processing and Decision Making*, 129–74. Chicago: University of Chicago Press, 1998.

——— and John M. Ratcliffe, eds. *Cognitive Ecology II*. Chicago: University of Chicago Press, 2009.

Enquist, Magnus, and Anthony Arak. "Neural Representation and the Evolution of Signal Form." In *Cognitive Ecology*, edited by Reuven Dukas, 21–87. Chicago: University of Chicago Press, 1998.

Finkelpearl, P. J. "Davies, John (1564/5–1618), poet and writing-master." In *Oxford Dictionary of National Biography*. Oxford University Press, September 2004. DOI: https://doi.org/10.1093/ref:odnb/7244.

Greenberg, Paul. *Four Fish: The Future of the Last Wild Food*. New York: Penguin, 2010.

Herendeen, W. H. "The Rhetoric of Rivers: The River and the Pursuit of Knowledge." *Studies in Philology* 78, no. 2 (Spring 1981): 107–27.

Hopkins, Gerard Manley. *Poems and Prose*. Edited by W. H. Gardner. 1953. Reprint, London: Penguin, 1985.

———. *The Collected Works of Gerard Manley Hopkins*. Vol. 3: Diaries, Journals, and Notebooks. Edited by Lesley Higgins. Oxford: Oxford University Press, 2015.

Howlett, Robert. *The Angler's Sure Guide*. London, 1706.

Kilgour, Maggie. "Writing on Water." *English Literary Renaissance* 29 (1999): 282–305.

Lakoff, George, and Mark Johnson. *Metaphors We Live By*. Chicago: University of Chicago Press, 1980.

Leland, John. Kykneion asma. Cygnea cantio. London, 1545. STC (2nd ed.) 15444.

Lewis, Samuel. *The Book of English Rivers*. London: Longman, 1855.

McDonald, John. *The Origins of Angling*. New York: Doubleday, 1957.

McRae, Andrew. "Fluvial Nation: Rivers, Mobility and Poetry in Early Modern England." *English Literary Renaissance* 38 (2008): 506–34.

[Samuel, William]. *The Arte of Angling*. 1577. Edited by Gerald Eades Bentley. Princeton, NJ: Princeton University Press, 1956.

Smith, Nicholas D. "Dennys, John (*d.* 1609), poet," In *Oxford Dictionary of National Biography*. Oxford University Press, September 2004. DOI: https://doi.org/10.1093/ref:odnb/7510.

*This present boke shewyth the manere of of hawkynge [and] huntynge and also of diuysynge of cote armours* ... [The Book of Saint Albans]. Westminster, 1496. STC (2nd ed.) 3309.

*Treatyse of fysshynge wyth an Angle*. In *This present boke shewyth the manere of of hawkynge [and] huntynge and also of diuysynge of cote armours* ... [The Book of Saint Albans]. Westminster, 1496. STC (2nd ed.) 3309.

Vallans, William. A tale of two swannes Wherein is comprehended the original and increase of the riuer Lee commonly called Ware-riuer: together, with the antiquitie of sundrie places and townes seated vpon the same. London, 1590. STC (2nd ed.) 24590.

Walton, Izaak. *The Compleat Angler or, the Contemplative Man's Recreation*. 1st ed. London, 1653. Wing (2nd ed.) W661.

Woolf, Virginia. *A Room of One's Own*. 1929. Reprint, New York: Harcourt, 1989.

# 4 Shakespeare's Angling Devices in *Antony and Cleopatra*

A philological study of *Antony and Cleopatra* has to reckon with the formidable chain of storytellers that link Shakespeare to his characters. The piscatory hobby of the two lovers cannot be the distinctly English sport discussed in the previous chapter, but it does come down to the playwright through a vernacular rendering of a classical text: *Lives of the Noble Grecians and Romanes*, Thomas North's translation of Plutarch, published in 1579.[1] In a treatment of the complex relationships between Shakespearean plays and their Plutarchian origins, John Denton exhorts readers to "bear in mind that Shakespeare was reading a domesticated contemporary translation ... and not Plutarch's Greek text or even a Renaissance Latin translation."[2] Denton is hesitant to accept the claim that "Shakespeare can on the odd occasion appear to be closer to Plutarch's original Greek text than to those of his 16th century vernacular translator(s)," since there is ample evidence, in plays that draw from Plutarch's *Lives*, "of Shakespeare's almost exclusive reliance on North."[3]

In order to properly hear the language of *Antony and Cleopatra*, it makes sense to treat both Shakespeare's play and North's prose as efforts of translation and adaptation, interesting in their fidelities to their respective sources and in their departures from them. My analysis of the angling trope in *Antony and Cleopatra* thus attends at times to the French of North's immediate source, Jacques Amyot's translation of Plutarch (*Les Vies des hommes illustres, grecs et romains*).[4] Although Shakespeare streamlines North's language in a way that occasionally mimics the less varied diction of Amyot, such coincidences of style are not presented here as evidence of any direct familiarity with Amyot's text on Shakespeare's part. Rather, these doublings-back show that the playwright and the English translator before him were similarly guided by the demands of their respective genres and potential audiences. The important point is that at least three previous iterations of the famous lovers—with their fishing tackle—stand behind the couple we meet on the seventeenth-century English stage.

In framing the character of Antony, Shakespeare draws from North's Plutarch the to-and-fro movement between the triumvir's remarkable endurance of physical hardship and his stubborn preference for leisure. The source material provides many examples of each, and Shakespeare's play, in turn, opens with several acknowledgments that the general has set aside his laborious martial duty in favor of fun. Antony's characterization, in the early scenes of the play, depends upon a particular understanding of recreational time that gets its fullest expression in accounts of angling. The sportive spirit of Antony and Cleopatra is best observed on the verges of rivers, and is exemplified by the totally unnecessary activity of freshwater fishing.

A strong distinction between riparian and oceanic waters organizes the political story of the play, separating Antony fundamentally from "the empire of the sea" commanded by Sextus Pompeius (1.2.161).[5] The contrast between the open sea and the river's mingling of land and liquid is called up again and again in the language of *Antony and Cleopatra*, building toward the final maritime disaster that signals the end of the general's career. Shakespeare makes much of his source's anecdotal references to fishing, letting the piscatory theme radiate outward. We find it echoed in the emphatic distinction between fluvial and pelagic environments, and in repeated references to *sport* as a category of human behavior and feeling. In the playwright's reading, the mention of Antony and Cleopatra's fishing habits must have resonated with the less intimate narrative of the general's political and martial downfall. If one thing is certain about Antony at the beginning of the play, it's that he has gone fishin'.

## Sport

Antony's vacillation between gutsy leadership and luxurious withdrawal from matters of state is emphasized in North's rendering of Plutarch, where the general "[gives] him selfe againe to his former riot and excesse" (979), "delighting in … fond and childish pastimes" (983). According to North's text, the common people came not only to "mislike" Antonius, "but also hate him for his naughty life: for they did abhor his banckets & drunken feasts he made at vnseasonable times" (974). The general's error is to miscalculate the commands of the time, squandering opportunities for careful political action by losing himself in fruitless activities. North calls such pursuits "sports," and cautiously avoids repeating Antony and Cleopatra's error of self-centered indulgence: rather than providing a lengthy catalog of the couple's activities, he limits himself to the description of just one. His caution is an echo of his immediate source, the narrative voice of Jacques Amyot's French Plutarch, which determines, "Si seroit trop grande simplesse de vouloir ramasser & reciter tous les esbattemens qu'ilz feirent lors en se jouant: mais j'en racompteray seulement un entre les autres: Il se meit quelquefois à pescher à la ligne …" (630r). "But to reckon vp all the foolishe sportes they made, reuelling in this sort: it were too fond a parte of me," says North's narrator, "and therefore I will only tell you one among the rest. On a time he went to angle for fish ..." (983).

Following Plutarch, both the French translator and the English after him make a point of selecting a single example from the available entertainments. All three narrators demonstrate their careful husbanding of storytelling time by reminding the reader of the narrative filter that catches only the most relevant information and lets all the excess pass through. In North's prose, the difference between the temperate narrator and the hedonistic lovers is sharpened by the dangling modifier "revelling in this sort." Hanging between the historical activities of the couple and the potential imprudence of the speaker, the phrase can equally refer to either. It also rhymes with the "sports" just mentioned, and prepares the reader to acknowledge the sport-sorting that has identified a solitary, telling example from "among the rest." The synonyms *foolish* and *fond*, correspondent in both sound and meaning, provide further proof that our narrator is anxious to be distinguished from his subject. We will return to this best-possible anecdote later.

Shakespeare picks up both the generally fraught characterization of leisure and the specific language of recreation found in North's life of Antony. The word *sport* is spoken five times in the course of the play, with occurrences distributed across the action and among the characters. Twice it comes from Antony himself, both times with reference to a range of potential recreational pursuits. In his first appearance, the general asks his lover, "What sport tonight?" (1.1.49). The question has the kind of ambiguity that suggests sexual play above all else, much as Antony's deliberately vague pronouns in this scene do: "The nobleness of life / Is to do *thus*; when such a mutual pair / And such a twain can do'*t*;" and "Last night you did desire *it*" (1.1.38–40, 57; emphasis added). Certainly sex is often what Shakespeare means by *sport*,[6] but Antony proposes an activity that is erotic in a different way: a spectator sport that requires an element of disguise. Against Cleopatra's insistence that they "hear the ambassadors" (1.1.50), Antony suggests that they venture out of the royal residence: "all alone / Tonight we'll wander through the streets and note / The qualities of people" (1.1.54-56).

Incidentally, this kind of people-watching can itself be compared to the sport of angling, since both pursuits entail a kind of visitation that takes enthusiasts out of their own habitat and places them in a novel environment. The urban rambler and the eager angler are similarly delighted by a vision of biotic diversity. In *The Compleat Angler*, Isaak Walton's expert fisherman tells his student that the joy of angling comes in part from observing a staggering array of species, many of which are truly bizarre: "These seem to be wonders, but have had so many confirmations from men of Learning and credit, that you need not doubt them; nor are the number, nor the various shapes of fishes, more strange or more fit for contemplation, than their different natures, inclinations and actions."[7]

Antony says he wants to be "all alone" with his lover, but surrounded by strangers. The couple will be curiously secluded in the crowded city, set apart in their own minds by their fundamental difference from those they watch. Like the cock- and quail-fighting mentioned later (2.3.34–35), angling and people-watching are sports that make other, less valued lives into a source of entertainment. Further, beyond the element of spectacle that characterizes these hobbies, there is a sense that pure luxury lies in doing for fun what others do out of necessity. Antony has no more need to gather details about ordinary people than he has to catch fish for his supper; the primary yield in both cases comes instead from the titillating paradox of a private activity that is performed out in the open. The pleasure of "wander[ing] through the streets" is derived in part from an element of disguise: the lovers will not sit at the center with all eyes on them, but will make a game of observation by going incognito. The gentle deception they perform by hiding themselves in public closely resembles the trained angler's stealth on the riverbank.

As we've seen, the narrative voices that bring Antony's story to Shakespeare are selective in their presentation of the general's activities. In their efforts to reveal but not to duplicate Antony's frivolity, the historical authorities of Plutarch, Amyot, and North are loath to squander precious time. Antony is, too, but his priorities are opposite: "Let's not confound the time with conference harsh; / There's not a minute of our lives should stretch / Without some pleasure now" (1.1.47–49). Love and sex, according to Antony and others, have renewed his commitment to leisure. Sport of any kind, being the antithesis of official business, operates as a celebration of the besotted "mutual pair," and the elasticity of time is to be exploited for their pleasure alone. This is the substance of the complaint with which Philo opens the play, Antony's turn from martial prowess to indoorsy dissipation:

> Nay, but this dotage of our General's
> O'erflows the measure. Those his goodly eyes,
> That o'er the files and musters of the war
> Have glowed like plated Mars, now bend, now turn
> The office and devotion of their view upon a tawny front.
> (1.1.1–6)

Tellingly, the contrast between Antony's laudable service as general and his conduct in Alexandria is also the first matter raised by Octavius Caesar, who seems especially offended by accounts of his fellow triumvir's angling: "From Alexandria / This is the news: he fishes, drinks, and wastes / The lamps of night in revel" (1.4.3–5). Unknowingly echoing Antony's language, where time is "confounded" by attention to politics, Caesar condemns the general's behavior as a failure to recognize what time it is. Such oblivious intemperance could be excused in a period of peace and stability, but right now Antony is shirking urgent responsibilities at home, "confound[ing] such time / That drums him from his sport, and speaks as loud / As his own state and ours" (1.4.28–30). Fishing, according to Caesar's formulation, is equal in virtue to drinking and staying up all night—it is an emblem of luxury and idleness, a mere *sport* whose yield is not sustenance but pleasure. Antony angles in a peaceful landscape utterly unlike the war-torn wilderness where, in Caesar's memory, the soldier was once sustained by what he could gather:

> Thy palate then did deign
> The roughest berry on the rudest hedge.
> Yea, like the stag when snow the pasture sheets,
> The barks of trees thou browsed.
> (1.4.63–66)

Giving this close paraphrase of North's Plutarch to Caesar shortly after the combined reference to fishing, drinking, and revelry, Shakespeare yokes the play's theme of feasting to the image of the angling lovers. While once he foraged for berries and bark at the edge of starvation, now Antony fishes for food he doesn't need, in a world where he doesn't belong.

Caesar's frustration is legitimized by evidence that he, too, participates in leisurely activities when there is no pressing matter of state calling him away. The Soothsayer speaks to Antony of Caesar's greater "natural luck" (2.3.24), and leaves Antony to ponder their shared recreational history:

> Be it art or hap,
> He hath spoken true. The very dice obey him,
> And in our sports my better cunning faints,
> Under his chance.
> (2.3.30–33)

These sports include cock-fighting and quail-fighting, contests that draw other creatures' bodies into conflict with one another. Early modern audiences would be likely to see such activities as comparable to sportfishing. Tips on cock-fighting and angling are combined, for example, in Gervase Markham's 1614 manual *The Pleasures of Princes*, and the play puts both sports in the broader category of gambling.[8] However, Caesar's disdain of fishing apparently does not extend to other recreations that depend so heavily upon the manipulation of animal behavior. Beyond his own implicit preference, there are additional reasons to regard angling as distinct and non-Roman. In the sporting anecdote selected by Plutarch to illustrate Antony and Cleopatra's frivolity, fishing is not a game of chance but a skilled pursuit. Antony lacks both the practice and the patience to succeed at it, but his lover is a master of the angle.

## The Fisher Queen

The angling story is related in a scene that is centrally concerned with ways of passing the time. It is framed, appropriately, by Cleopatra's contemplation of leisure. As she awaits news from Antony, the queen entertains several ideas about potential activities: she calls for music, considers playing billiards, and engages the eunuch Mardian in a game of bawdy wit. Ultimately, she sends for her fishing rod. Imagining the sport, she compares her catch to the man who has slipped away:

> Give me mine angle. We'll to th'river. There,
> My music playing far off, I will betray
> Tawny-finned fishes. My bended hook shall pierce
> Their slimy jaws, and as I draw them up,
> I'll think them every one an Antony,
> And say 'Ah ha! you're caught.'
>
> (2.5.10–15)

The sliminess of these personified fish, presumably a challenge for the hopeful angler, recalls riparian imagery from other speakers. When they talk about Pompey's encroachment, both Antony and Caesar use wet life forms as figures for the shifting loyalties of the populace; their sketches of fish and rushes participate in a broad pattern of riverbank images. Antony complains of "Our slippery people, / Whose love is never linked to the deserver / Till his deserts are past" (1.2.161–63). The subtle reference to fish is here enforced with the verb *to link*. A loop in a horsehair fishing line was called a *link*.[9]

To Caesar, the "common body" is like another slimy being, "a vagabond flag upon the stream, / [that] Goes to, and back, licking the varying tide, / To rot itself with motion" (1.4.44–47). The flag, best understood in early modern English as a reed or rush, is a plant that thrives in an estuarine zone but can have its structural integrity compromised by drastic fluctuations in the tide.[10] Flags are associated with both the habitat and flesh of freshwater fish in the piscatory literature. The instructor in William Samuel's *Arte of Angling* directs his student to harvest bait for the bream from the wet roots of such plants: "You must pull up flags by the roots out of the water, and in the roots you shall find white worms as big as gentles, and they be very good."[11] Leonard Mascall's treatise on angling and pisciculture takes it for granted that flags will spring up voluntarily in any pond, and will need regular weeding so they don't spoil the fish:

> If you will haue profite of your fish, in your pondes and pooles, ye must haue a care alwayes to clense them from three yeare to three yeare, in taking away all weedes, rushes, and flagges, for they doe

> greatly stuffe and trouble the fish, and makes them to be more slymie, and of a worser taste.[12]

Mascall also mentions that many people are turned off by the texture of eel flesh, finding it to be "a flaggie and slymie meate."[13] Flags, then, are associated with a sliminess that is thought to come from an overactive appetite and a lack of vigilance. This image, in turn, is echoed by the servants who marvel at the rapid intoxication of the banquet guests in the second act. Like the flag in Caesar's simile, the revelers' verticality and self-determination is overcome by a superfluity of liquid: "Some o'their plants are ill rooted already; the least wind i'th'world will blow them down" (2.7.1–2). While these botanical life forms are prone to dangerous slippage because they are not fully aquatic, the slimy horizontality of Cleopatra's fish is a protective adaptation. They can get away and live their lives—unless, of course, they're hooked.

The queen's proud description of her "bended hook" appears redundant, since the defining feature of the device is its curvature. We know from angling manuals reaching back to the *Treatyse* that early hooks were merely manipulated needles, appropriated from their intended use and literally bent into service. The *Treatyse* instructs its reader to heat the needle and "bende hym lyke to the bend fyguryd hereafter in example," referring to an illustration of hooks in various sizes which appears several pages later (see Figure 1.1).[14] The original meaning of *angle*, as applied to fishing, was not the rod but the hook itself, aptly named after its characteristic bend.[15] Why does Shakespeare fill out the line with "bended," then? The phrase presents a tidy parallel with what follows, the queen's "bended hook" (an extension of her dry human body) being in direct contention with the fish's "slimy jaws." It also picks up the language Philo has used to criticize Antony's new priorities: "those his goodly eyes … now *bend*, now turn / The office and devotion of their view upon a tawny front" (1.1.2–6, emphasis added). In a further echo, Cleopatra here describes the color of the fish as "tawny-finned," using Philo's very word for *her* skin. Even as she compares Antony to a clutch of captured fish, the play's lexical repetition holds her up as a bright brown, semi-aquatic body. A shiny baited hook is a bent offering that can turn a creature around, convince it to abandon its path, and accept the risk of being taken. Without seeming to realize it, Cleopatra points up both the allure of her own shimmering skin and the vulnerability of a lovestruck body by admiring the color of her imaginary catch.

There is also a possibility that the bend to which the queen refers is not the broad curvature of the hook but the smaller, sharper process within it: the barb. The word most often applied to this protrusion in Shakespeare's time was *beard*. Although the two words clearly come from the same bodily metaphor, and both were used for backward-curving jabbers on pointed instruments, the inner point on a fish-hook was not called a *barb* until much later.[16] In the *Treatyse*, the beard is fashioned first, before the decisive rounding of the hook: "reyse the berde wyth your knyfe and make the poynt sharpe."[17] According to Dennys's *Secrets of Angling*, hooks must be "wrought smooth and cleane withouten cracke or knot / And bearded like the wild Arabian goat" (B4r). If *bended* is a likely compositor's error for *bearded*, it could be that the first Shakespearean Cleopatra spoke of the barb itself rather than the general bend of the hook—not just the shape that allows her to pull the fish toward her, but the more subtle protrusion that can't be removed without tearing the creature's flesh. Whether it is explicitly named or not, this tiny beard is the element of the queen's fishing apparatus that ultimately allows her to "draw them up" and call them "caught." In a play where anxiety and titillation swirl around the reversal and unraveling of gender, Caesar's fear that Antony is "not more manlike / Than Cleopatra" (1.4.5–6) receives a barbed response in the image of a triumphant fishing queen.

With her stated intent to "betray" the fish in the nearest river, Cleopatra may seem to be thinking only of Antony, and not of fishing at all. It makes sense for the impatient lover to entertain fantasies of beating her beloved to the punch, of making him seem ridiculous before he has a chance to embarrass her. Feeling caught herself, she might imagine a final act of possession that brings Antony back to her against his will. It is certainly possible to read backwards to this moment from the end of the play and recognize in the queen's phrasing a premonition of her later machinations. But *betray* is more fascinating for its origins in the practical language of angling, where catching fish had long been represented as an art of deception. What appears to be an anthropomorphizing metaphor is actually a literal confirmation of what angling so often requires: the representation of one thing as something else entirely. Cleopatra sounds like the piscatory authority in the medieval *Treatyse*, who assures his reader that "it shall be to you a very pleasure to se the fayr bryght shynynge scalyd fysshes dysceyued [deceived] by your crafty meanes and drawen vpon londe."[18] The instructional voice in Dennys's *Secrets of Angling* explains that a hopeful sportfisher might go home with no catch whatsoever, in spite of having all the gadgets in the world, "if he know not how to vse the same, / And with what baites to make the fishes fooles."[19] When Donne's amorous speaker in "The Bait" refers to the treachery, betrayal, and deceit inherent in angling,[20] he simply echoes—as Cleopatra does here—the conventional language of a sport that relies upon outright trickery. A practical joke played at Antony's expense will soon become the focus of the queen's reminiscences, but she is focused, at the start, on her ability to make fools of fish.

As she fantasizes about her piscatory success, Cleopatra is reminded of her actual superiority with rod and line. Shakespeare gives the angling anecdote from Plutarch to the attendant Charmian, who recollects the episode with glee:

> 'Twas merry when
> You wagered on your angling, when your diver
> Did hang a salt fish on his hook, which he
> With fervency drew up.
>
> (2.5.15–18)

This version of the story is a radical abbreviation, equally interesting in its inclusions and omissions. North closely follows Amyot in delivering a narrative with several stages, but Shakespeare pares the plot down to a single prank. In Plutarch and his translators, Antony is frustrated by his failure as an angler. Wanting to impress Cleopatra, he enjoins several nearby fishermen to dive into the water with fish they've already caught and attach them to his hook. In this way, the general seems to catch two or three fish. His lover recognizes the entire scheme, but feigns wonder at his apparent skill. The next day, to expose the trick, Cleopatra arranges for a diver to hook "some old salte fish" onto Antony's line. She also ensures that a large audience will be there to partake of the spectacle. Antony excitedly "snatche[s] vp his line" and is roundly mocked. The punchline of the story is Cleopatra's consolation: "leaue vs (my Lord) AEGYPTIANS (which dwell in the contry of PHARVS and CANOBVS) your angling rodde: this is not thy profession: thou must hunt after conquering of realmes and contries" (North, 983).

North retains the metaphor of the hunt from Amyot's Cleopatra: "ta *chasse* est de prendre & conquerir" (639v, emphasis added). The expression shows that angling is understood to be an aquatic hunt, a sport among sports. It also suggests that Antony is fundamentally not suited to recreation, but instead belongs on military campaigns and at the helm of a growing state. In Shakespeare, there is no such culminating statement about Antony's calling. As the queen awaits news of her beloved, knowing he has been duly called away by his imperial obligations, she reflects not on his political duty but on the jollity and variety of their mutual recreation. If she can't fish *with* him, she will have to fish *for* him. Earlier in the play, when she tries to get Antony to stick to business before retreating to pleasure— "Hear the ambassadors" (1.1.50)—she sounds like the Plutarchian queen who redirects the focus of her lover from the frivolous passing of time to the triumphant conquering of lands. But the fishing story won't be reduced (or elevated) in the play to any single, sober exemplum. Instead, it is part of a suite of memories that represent, for Cleopatra, better times—a season of teasing that is quickly receding into the past.

In Amyot, both the angling itself and the winsome salt-fish plot are called "esbattements," a word North translates as "foolishe sportes" (Amyot, 639r; North, 983). Amyot uses a related term, "esbats," to tie up the episode and move into the next phase of the historical account. North repeats "fond" from his very deliberate introduction of the anecdote; there, he was at pains not to appear "too fond," and here, at the story's conclusion, he marks the contrast between Antony's amusement and the impending arrival of bad news by emphasizing the pleasure the general takes in "fond and childish pastimes" (983). The effect of both renderings is to conflate the sport of angling and the practical joke that follows, including Cleopatra's teasing affirmation of Antony. It isn't clear, after the queen's conciliatory statement, whether mere fishing or the whole game of harmless deceptions is what constitutes the best illustration of the couple's leisurely conduct. Shakespeare does not exactly remove this ambiguity by abbreviating the tale, but he alters the story by omitting its moral. If this Cleopatra once made an emblematic lesson of her lover's vanity, the fishing high jinks nevertheless live in her memory only as a funny, sexy prank, not as an opportunity to reflect upon Antony's true military calling.

In response to Charmian's reminder about the salt-fish ploy, Cleopatra recollects the more private activities that necessarily followed the angling gambit:

> I laughed him out of patience, and that night
> I laughed him into patience, and next morn,
> Ere the ninth hour, I drunk him to his bed,
> Then put my tires and mantles on him whilst
> I wore his sword Philippan.
>
> (2.5.19–23)

The queen's prepositions indicate deliberate, guided movement aided by humor and drink: she "laughed him *out of* [and] *into* patience" "drunk him *to* his bed," and, if the game of dress-up was not consensual, played another trick on him. Borrowing his sword while he's passed out, Cleopatra inverts the image of the rigid salt fish that is secretly attached to Antony's line. To different degrees and by different means, both of these phallic accessories come to represent the fragility of reputation. The queen's playful swapping of gear marks a middle term between Antony's minor humiliation at hauling up a cured fish and the pitiful commandeering of his sword as evidence of his demise (4.15.110–11). When the sword is taken up by a disloyal Dercetus, the poignant spectacle of the general's dying body recalls Cleopatra's precious memory of that same body in a contrived state of vulnerability. On the riverbank and at the bedside, she temporarily alters her lover's image for the sake of fun. Her play might, like any good teasing, have affection and admiration at its core. The dress-up episode takes the place, in strict sequential terms, of the deferential moral Cleopatra utters in Shakespeare's source. It may also serve a similar function to that utterance, of revealing Antony's past (and potential) strength through his present weakness.

## Drawing Up

Forms of the verb *tirer* are applied consistently to the sport's gestures in the translation by Amyot ("retira ainsi deux ou trois fois sa ligne avec prise" and "tira incontinent sa ligne" [639r, v]), but North varies his diction to give us "snatched vp" and "brought vp" (983). The same pattern applies in the scene of Antony's death: Amyot repeats *tirer* throughout, while North introduces "trised … vp," "plucked vp," "raised vp," "lift[ed] … vp," and "gotten … in" to describe the various motions that elevate Antony to the window and pull him in (1006). Shakespeare, in compressing the action, also limits the lexical variety in his images of hoisted bodies. Like Amyot, he borrows the language of Antony's fatal ascent from an earlier description of fish being pulled from the water. Cleopatra calls her whole company and the guard below to the task of heaving the beloved body up: "Help, Charmian, / Help, Iras, help, help, friends below! / Let's draw him hither" (4.16.12–14). She then repeats the command, addressing Antony himself: "we must draw thee up" (4.16.31). The consistency of this diction, as other readers have noted, reinforces the piscatory quality of the death scene. Significantly, *draw* is used at this moment in the action only by Cleopatra, the play's consummate angler. If the comparison is yet too subtle to be recognized, the queen confirms it with a further echo: "Here's sport indeed!" (4.16.33). The usage resonates ironically with Antony's "What sport tonight?" and with all that we know of the couple's leisurely pursuits. The erotic charge that Cleopatra has felt

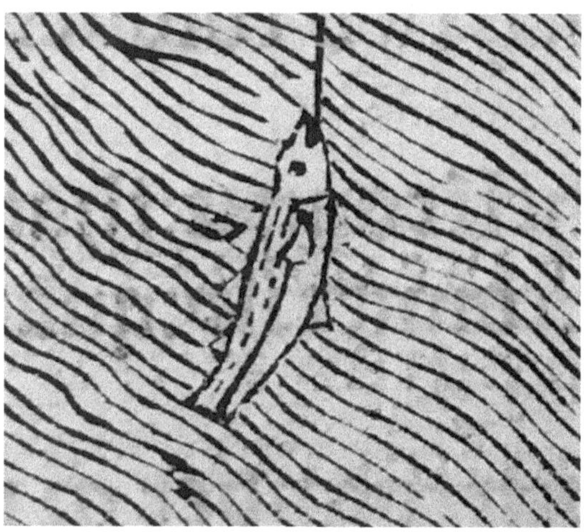

*Figure 4.1* Detail from the frontispiece of the *Treatyse of fysshynge wyth an Angle*. Beinecke Rare Book and Manuscript Library, Yale University.

*Figure 4.2* Detail from the frontispiece of *The Secrets of Angling*. Folger Shakespeare Library.

under Antony's weight is tragically reversed, as she must now lift him up to die on a sexless bed.

Noting that Antony "does not fall as convention dictates but rises to die," Leslie Thomson explains that the death scene "is prepared for both visually and verbally by what comes before it," and that it is "the last of a series in which Cleopatra causes Antony to move towards her and he willingly does so. In fact, Cleopatra's power to pull Antony to her, both physically and emotionally, can be seen as a key organizing principle of the language and structure of the play."[21] Thomson observes that the repeated *drawing* in the dialogue of the death scene echoes angling terminology used by Cleopatra and Charmian, and finds suggestive occurrences of the verb elsewhere in the play.[22] The choreographic awkwardness of the action is, for Thomson, deliberately construed by Shakespeare to accord with tensions already at work in the play: "Oscillation, equivocation,

*Figure 4.3* Detail from the frontispiece of *The Compleat Angler*. Beinecke Rare Book and Manuscript Library, Yale University.

and ambiguity characterize both the actions of Antony and Cleopatra and our response to them. This disconcerting quality is made particularly apparent by the staging of 4.16."[23]

Looking back at several decades of stage history, Marvin Rosenberg observes that the difficulty of hoisting Antony up has been "dodged" by various means: the use of a block and tackle, for example; "or with a safe halter round his waist, he has seemed to be holding on with one hand to a rope that raised him. Or he has been lifted, like a package, in a chair, or sling, or net, all his special aura and dignity gone."[24] For Rosenberg, these alternatives run against Shakespeare's intent by foregrounding the infelicity of action that seems both inevitable and appropriate to Thomson. But if Antony is bundled and hauled aloft "like a package," we have onstage an image that accords with Shakespeare's source: a body "trussed," in North's phrasing (1006), or "empacqueta" in Amyot's (654v). The language of angling, so pointedly echoed by Cleopatra in the repetition of *draw* and the bitter evocation of *sport*, is borne out in the stage business itself. Whatever lines are let down, Antony must be fastened to them somehow. If they become a net in the process, the action finally presents an image that has twice been conjured in the play's dialogue: a captured body, already defeated, being hauled ashore.[25]

Because there is no reference to the technology of the reel until Thomas Barker's angling treatise of 1651,[26] the gear Shakespeare's audience would have associated with landing freshwater fish necessarily included a long-handled net. John Dennys stipulates that one needs "A little Net that on a Pole shall stand / The mighty Pike or heauy Carp to Land."[27] With an eye to conservation, Leonard Mascall cautions readers about the irresponsible use of fine-mesh nets, which are all too effective at catching fish that need more time to mature:

> [Such nets] are destroyers of fish before they are growen to any bignesse. These are not meete to be vsed but of certaine Gentlemen in their seuerall waters, I would wish no running waters should be let to any fisher man, without order what mesh, what nets, he or they shall in what moneths of the yeare to refraine fishing, vpon paine to forfaite his lease and all such engins.[28]

Of course, nets were not used only in the capture of fish; Mascall himself goes on to include models for several bird-trapping nets.[29] In Shakespeare, where fishing is so rarely mentioned, characters are much more likely to refer to the nets used to capture avian and terrestrial animals.[30] The exception is not an aquatic creature but the "rusty armour" accidentally caught by the commercial fishermen of *Pericles*. Although these laborers are opposite to Cleopatra in their fishing—on the open sea rather than the riverbank, for much-needed pay rather than mere fun, and with large nets rather than a single rod—both the dialogue and the stage directions echo her piscatory diction. A moment after the Second Fisherman declares that he will "go draw up the net" (2.1.84), he and his fellow are seen doing just that: "*Enter [Second and Third] Fishermen, drawing up a net*" (2.1.103.1).

Despite Rosenberg's concern that the netting of Antony could strip him of his dignity, that particular staging is consistent with Shakespeare's figuration of nets as traps that reveal their contents. In *Henry V*, the Archbishop of Canterbury says the Salic law impeding Henry's claim to France is a net in which the French monarchs foolishly "choose to hide" (1.2.93). When Iago plots to fabricate from Desdemona's virtue "the net / That shall enmesh them all" (2.3.323–24), he promises to produce not only the downfall of his victims but the spectacle of their fatally entangled conflict. The apparatus with which Antony is elevated, whatever its precise form, is netlike in its ability to keep the dying general before our eyes even as it transports him. Observed from above, below, and beyond—by Cleopatra and her women, his guard, and the audience in the theatre—Antony shows, like a fish drawn up from the water, that the moment of his death has not yet come.

The playwright may also have had in mind another typical usage of the verb *to draw*, one that is compatible with the imagery of angling but more precisely related to the labor of literary adaptation. The phrase *to draw up* was used in early modern recipe books to refer to the use of a strainer in bringing food "to the proper consistence."[31] Like a net, a strainer can separate the desired from the extraneous, the solid from the liquid, the fish from the water. When Shakespeare draws certain elements up from his source material, he follows the self-conscious example of North (who followed Amyot, who followed Plutarch), choosing the matter that is most consistent with the characters he wishes to reveal. Strainers and nets behave like the apparent selectiveness of the Plutarchian narrator, drawing one substantial specimen from the stream of available stories. In *The Winter's Tale*, a play that repeatedly deploys angling metaphors, an unnamed character seems to be thinking of nets when he describes the arresting effects of a poignant spectacle. Recounting his own experience of seeing Perdita reunited with her penitent father, the Third Gentleman says that the proceeding was "pretty," that it "angled for mine eyes," and that it "caught the water, though not the fish" (5.2.64–66). The challenge of appealing to piscine vision—of catching the fish's eye—is a central concern of anglers, as we've seen. It is not surprising that this speaker should use the sport to represent the seizing of creaturely attention, particularly through the eyes. But the Third Gentleman's metaphor goes further, conjuring up a net that is either magical or completely ineffective: it lets the eyes get away, and captures only their tears.[32] When Antony is lifted up to his final rest, the contraption that carries him takes the audience's eyes with it, at least temporarily. If the spectacle provokes weeping, it confirms Cleopatra's power as an angler who catches both fish and water. She brings Antony up to the level of tragedy.

Rather than relying on the dialogue to communicate the tremendous physical effort involved in bringing Antony into Cleopatra's embrace—North has the queen "stowping downe with her head, putting to all her strength to her vttermost power" (1006)—Shakespeare lets the labor unfold in real time before the audience. The bodily struggle is a physical consequence of the intellectual and emotional manipulation that has contributed to Antony's tragic position. It is Cleopatra's false message of her suicide, followed by its belated retraction, that most directly draws his enfeebled body to the monument. The stage business in 4.16 operates as an embodiment and culmination of the queen's overwhelming emotional draw, as Thomson suggests. It also recalls the barge scene that Shakespeare meticulously cribs from North and delivers through the admiring voice of Enobarbus. The couple's first encounter, like their last, invites comparisons between Cleopatra's favorite sport and her ability to get people right where she wants them.

## Down by the Riverside

Antony and Cleopatra angle for each other in the narrative of their first meeting, which begins with a volleying of messages across the border between land and water. The queen drifts on her opulent barge toward the city on the banks of the river Cydnus, and the bewitching fragrance of the vessel "hits the sense / Of the adjacent wharfs" (2.2.211–12), drawing people down to the water's edge. Antony resists the pull, stays put in the marketplace, and "sen[ds] to her" (2.2.218) when she lands. His invitation is countered with Cleopatra's—"It should be better he became her guest" (2.2.220)—and Antony is caught the moment he sees his hostess. Like a hungry fish who is destined to be consumed, he "goes to the feast, / And for his ordinary pays his heart / For what his eyes eat only" (2.2.223–25). The queen's invisible, inexorable hook has its outward manifestation in the floating stage that glitters on the Cydnus, decorated with the balletic "bends" of her gentlewomen (2.2.207). Their motions, as they "tend" their mistress "i'th'eyes" (2.2.206), match Philo's account of Antony's enchantment from the beginning of the play, and share an orientation with the general's attentive gaze (the eyes that "bend" their view toward Cleopatra). While Enobarbus's barge scene closely follows the diction in North's description, the *bends* of the gentlewomen are introduced by Shakespeare. It is appropriate that *bending* should return in Cleopatra's talk of fish hooks a few scenes later. Her piscatory know-how is of a piece with the gleaming spectacle of the barge; to catch fish and people alike, the angler appeals to external appetites through elaborate optics.

The Egyptian queen's orchestrated movement of human bodies, by means of both political finesse and erotic allure, is related to the fluctuation of waters in the play's riparian imagery. In the first moments of the action, Philo implicitly likens the famously flooding waters of the Nile to Antony's affection for Cleopatra, which "o'erflows the measure" (1.1.2).[33] Soon after, in echoing language that is itself an overabundance, the two lovers envision riparian floods. Antony's playful dismissal of news from home ("Let Rome in Tiber melt" [1.1.35]) returns as Cleopatra's curse of her own land, after she has heard of her lover's marriage to Octavia: "Melt Egypt into Nile" (2.5.78). When she later beholds her dying lover from above, Cleopatra tells the sun to forsake "the varying shore o'th'world" (4.16.11). Like the "varying tide" in Caesar's simile of the "vagabond flag," her diction once again calls up shifting verges, places where the elements meet and intermingle.

Now stuck in muddy grief, the queen sees the erosion and dissolution of even the most exalted life: at the moment of Antony's death, she says, "The crown o'th'earth doth melt" (4.16.65). The expression obviously renders Antony a supreme king, but it also envisions him as a high spot of ground, like the make-believe "crown of the cliff" from which Gloucester falls in *King Lear* (4.6.67).[34] In the repeated imagery of melting earth we find the love of the central couple being compared to the riparian zone that fosters it, a mixture that is sometimes fruitful but always precarious, threatening to dissolve both powerful identities and leave the lovers no solid ground. It is angling, more than any other activity, that brings Antony and Cleopatra side by side on the riverbank in our imaginations. By dredging up Plutarch's reference to their sport, Shakespeare gives the lovers a foundation that is tragically prone to erosion.

## Notes

1 Plutarch, *The lives of the noble Grecians and Romanes compared together by that graue learned philosopher and historiographer, Plutarke of Chaeronea*, trans. Thomas North (1579, STC [2nd ed.] 20066. Text citations of North refer to page numbers in this edition.

2 John Denton, "Translation and Manipulation in Renaissance England," supplementary volume, *Journal of Early Modern Studies* 1 (2016), 52.

3 Denton, "Translation and Manipulation," 63.

4 Plutarque, *Les Vies des hommes illustres, grecs et romains, comparées l'une avec l'autre, par Plutarque de Chaeronée*, trans. Jacques Amyot (1565), accessed April 18, 2018, http://gallica.bnf.fr/ark:/12148/bpt6k857106b. Text citations of Amyot refer to page numbers in this, the second edition. North apparently used the third edition, which was printed in 1574 at Lausanne (Tom Lockwood, "North, Sir Thomas [1535–1603?], translator," *DNB* [January 2008], https://doi.org/10.1093/ref:odnb/20315).

5 All text citations of Shakespeare's plays refer to act, scene, and line numbers from *The New Oxford Shakespeare: Modern Critical Edition*, eds. Gary Taylor et al., online edition (Oxford: Oxford University Press, 2016), DOI: 10.1093/actrade/9780199591152.book.1.

6 Stephen Partridge, *Shakespeare's Bawdy*, 3rd ed., s.vv. "sport, n.," "sport, v.," and "sportive" (1968; repr., Abingdon, UK: Routledge, 2001), 244.

7 Izaak Walton, *The Compleat Angler* (1653, Wing [2nd ed.] W661), 21.

8 Linda Woodbridge, arguing that "gambling is deeply ingrained in the play's culture," cites bird-fighting and fishing side by side in a list of Antony's gambling hobbies ("He beats thee 'gainst the odds': Gambling, Risk Management, and *Antony and Cleopatra*,") in *Antony and Cleopatra: New Critical Essays*, ed. Sara Munson Deats (New York: Routledge, 2005), 193.

9 *OED*, s.v. "link, n.2," sense 2a, accessed April 18, 2018, www.oed.com/view/Entry/108729?rskey=TTciPz&result=2&isAdvanced=false.

10 *OED*, s.v. "flag, n.1," accessed April 18, 2018, www.oed.com/view/Entry/70889?rskey=dfm6SW&result=1&isAdvanced=false.

11 [William Samuel], *The Arte of Angling* (1577), ed. Gerald Eades Bentley (Princeton, NJ: Princeton University Press, 1956), 57.

12 Leonard Mascall, *A Booke of fishing with Hooke & Line, and of all other instruments thereunto belonging* (1590, STC [2nd ed.] 17572), 34–35.

13 Mascall, *A Booke of fishing*, 33.

14 *Treatyse of fysshynge wyth an Angle*, in *This present boke shewyth the manere of of hawkynge [and] huntynge and also of diuysynge of cote armours ... [The Book of Saint Albans]*, (1496, STC [2nd ed.] 3309), h3r, i3v.

15 *OED*, s.v. "angle, n.1," accessed April 18, 2018, www.oed.com/view/Entry/7541?rskey=UDgNgi&result=1&isAdvanced=false.

16 *OED*, s.v. "barb, n.1," sense II.8, accessed April 18, 2018, www.oed.com/view/Entry/15369?rskey=ehHLGd&result=2&isAdvanced=false.

17 *Treatyse*, h3r.

18 *Treatyse*, i4r.

19 John Dennys, *The Secrets of Angling* (1613, STC [2nd ed.] 6611), C3v.

20 John Donne, "The Bait," in *The Complete Poems of John Donne*, ed. Robin Robbins (Harlow, UK: Pearson Education, 2010); lines 19, 8, and 25.

21 Leslie Thomson, "*Antony and Cleopatra*, 4.16: 'A Heavy Sight'," *Shakespeare Survey* 41 (1989), 77–78.

22 Thomson, "*Antony and Cleopatra*, 4.16," 82.

23 Thomson, "*Antony and Cleopatra*, 4.16," 85.

24 Marvin Rosenberg, *The Masks of Anthony and Cleopatra* (Newark: University of Delaware Press, 2006), 383.

25 In a seminar on "Shakespearean Horizons" led by Patricia Badir at the 2015 meeting of the Shakespeare Association of America, Kay Stanton brought it to my attention that Antony is lifted up like a fish in a net.

26 Speaking of another gambling angler, Barker recounts the story of a man who laid a wager on his ability to catch a four-foot-long pike "within the space of one Moneth, with his Trouling-Rod." He succeeds not by *trawling* with a dragnet, but by using a *troll*, or reel: "The manner of his Trouling was, with a Hazell Rod of twelve foot long, with a Ring of Wyre in the top of his Rod, for his Line to runne thorow: within two foot of the bottome of the Rod there was a hole made, for to put in a winde, to turne with a barrel, to gather up his Line, and loose at his pleasure" (Thomas Barker, *The Art of Angling* [1653, Wing B783], 9). There is no extant copy of the 1651 edition of Barker's treatise. *OED*, s.v. "troll, n.1," sense 3, accessed April 19, 2018, www.oed.com/view/Entry/206613?rskey=GzIj94&result=1&isAdvanced=false.

27 Dennys, *The Secrets of Angling*, B5v.

28 Leonard Mascall, *A Booke of fishing*, 34.

29 The images of nets appear in the treatise that is attached, with continuing pagination, to Mascall's fishing manual: *A Booke of Engines and traps to take Polcats, Buzardes, Rattes, Mice and all other kindes of Vermine & Beasts whatsoeuer, most profitable for all Warriners, and such as delight in this kinde of sport and pastime*, 86–87, 93.

30 To cite just two such examples: In *Macbeth*, Lady MacDuff refers to the netting of birds (4.2.34); in *Henry VI, Part 3*, Northumberland speaks of "the cony struggl[ing] in the net" (1.4.62).

31 *OED*, s.v. "draw, n.," sense I.8 compared with sense III.51, and sense 4 of "to draw up" in the list of phrasal verbs; accessed April 19, 2018, www.oed.com/view/Entry/57534?rskey=ENE9Q6&result=2&isAdvanced=false.

32 In J. A. Shea's essay on the angling metaphors of *The Winter's Tale*, this episode shows how "theatre itself plays the angler" (143). Shea examines the play's relationship with another kind of angling—the trickery and thievery

laid out in the coney-catching literature of Robert Greene, Thomas Dekker, and Thomas Harman. See Shea, "Recasting 'Angling' in *The Winter's Tale*," in *Shakespeare's World of Words*, ed. Paul Yachnin (London: Bloomsbury, 2015).

33 Antony, for his part, speaks in more laudatory terms of the river's potential to flood. He describes the Nile's fertile banks, referring to the sun as "the fire / That quickens Nilus' slime" (1.3.68-69), and later tells the Romans about the fecundity that is dependent upon its flooding: "The higher Nilus swells, / The more it promises; as it ebbs, the seedsman / Upon the slime and ooze scatters his grain, / And shortly comes to harvest" (2.7.19–22).

34 *OED*, s.v. "crown, n.," sense II.24, accessed April 19, 2018, www.oed.com/view/Entry/45059?rskey=j5Yk6b&result=1&isAdvanced=false.

## References


Barker, Thomas. *The Art of Angling*. London, 1653. Wing B783.

Dennys, John. *The Secrets of Angling*. London, 1613. STC (2nd ed.) 6611.

Denton, John. "Translation and Manipulation in Renaissance England." Supplementary volume, *Journal of Early Modern Studies* 1 (2016).

Donne, John. *The Complete Poems of John Donne*, edited by Robin Robbins. Harlow, UK: Pearson Education, 2010.

Lockwood, Tom. "North, Sir Thomas (1535–1603?), translator." In Oxford Dictionary of National Biography, January 2008. DOI: https://doi.org/10.1093/ref:odnb/20315

Mascall, Leonard. *A Booke of fishing with Hooke & Line, and of all other instruments thereunto belonging*. London, 1590. STC (2nd ed.) 17572.

Partridge, Stephen. *Shakespeare's Bawdy*. 3rd ed. 1968. Reprint, Abingdon, UK: Routledge, 2001.

Plutarch. *The Lives of the Noble Grecians and Romanes Compared Together by that Graue Learned Philosopher and Historiographer, Plutarke of Chaeronea*. Translated by Thomas North. London, 1579. STC (2nd ed.) 20066.

Plutarque. *Les Vies des hommes illustres, grecs et romains, comparées l'une avec l'autre, par Plutarque de Chaeronée*, translated by Jacques Amyot. Paris, 1565. http://gallica.bnf.fr/ark:/12148/bpt6k857106b.

Rosenberg, Marvin. *The Masks of Anthony and Cleopatra*. Newark: University of Delaware Press, 2006.

Samuel, William. *The Arte of Angling*. 1577. Edited by Gerald Eades Bentley. Princeton, NJ: Princeton University Press, 1956.

Shakespeare, William. *The New Oxford Shakespeare: Modern Critical Edition*. Edited by Gary Taylor, John Jowett, Terri Bourus, and Gabriel Egan. Online edition. Oxford: Oxford University Press, 2016. DOI: 10.1093/actrade/9780199591152.book.1.

Shea, J. A. "Recasting 'Angling' in *The Winter's Tale*." In *Shakespeare's World of Words*, edited by Paul Yachnin, 125–46. London: Bloomsbury Arden Shakespeare, 2015.

*This present boke shewyth the manere of of hawkynge [and] huntynge and also of diuysynge of cote armours …* [The Book of Saint Albans]. Westminster, 1496. STC (2nd ed.) 3309.

Thomson, Leslie. "*Antony and Cleopatra*, 4.16: 'A Heavy Sight'." *Shakespeare Survey* 41 (1989): 77–90.

Walton, Izaak. *The Compleat Angler or, the Contemplative Man's Recreation*. 1st ed. London, 1653. Wing (2nd ed.) W661.

Woodbridge, Linda. "'He beats thee 'gainst the odds': Gambling, Risk Management, and *Antony and Cleopatra*." In *Antony and Cleopatra: New Critical Essays*, edited by Sara Munson Deats, 193–211. New York: Routledge, 2005.

# 5 The Interrupted Aquatic Hunt in Wroth's *Urania*

Lady Mary Wroth's *Urania* follows the conventions of early modern romance by frequently sending its characters off on hunting excursions, but the narrative preoccupation with field and stream is not merely a literary echoing of generic standards. The predominance of hunting as a courtly recreation, and the detail with which it is rendered in the *Urania*, come more directly from Wroth's lifelong familiarity with the sport. By all accounts, this was an interest she cultivated during her upbringing at the Sidney estate of Penshurst, throughout her marriage to Robert Wroth and her relationship with William Herbert, and beyond. For the Sidneys and the Wroths alike, blood sport was a central element of hospitality, a way for families to regularly entertain guests for days at a stretch. In the case of Mary's husband, Robert Wroth, the hunt was a locus of both personal pleasure and professional duty. His father, the elder Robert Wroth, held the position of Royal Forester. In this capacity, he was charged with "ensuring that the various keepers who worked under him kept track of deer killed, maintained the herd for the king's use and caught 'poachers,' kept cattle out of the forest, and maintained the forest laws."[1] After his death in 1606, these responsibilities fell to his son, who had married Mary Sidney in 1604.[2] The couple hosted King James at their primary residence, Loughton Hall, which was near the king's hunting grounds of Enfield Chase.[3]

Robert was clearly "an avid huntsman,"[4] and there is evidence that Mary, too, enjoyed the sport. Jonson's Epigram 105 includes hunting in a catalogue of her talents, declaring that Wroth is like a goddess when "arm'd to the chase;" only Diana "so hit, and hunted so."[5] She typically stayed at Loughton through the autumn hunting season each year, hosting members of both her own family and the royal family. It appears that she also held hunting parties at Penshurst after her husband's death. William Herbert, Wroth's lover and the father of two of her children, was also a keen hunter who sought and held several positions as keeper of royal forests. For Wroth's biographer Margaret Hannay, the writer's frequent references to the sport, along with these circumstances of aristocratic life, amount to ample proof that she had "a love of hunting."[6]

Less is known of Wroth's interest in the aquatic hunt, despite the fact that anglers appear several times in the *Urania*. In his poem celebrating the bounty of Penshurst, Jonson situates the estate in a riparian zone. The property has plenty of pasture for livestock in "the lower land, that to the riuer doth bend." That stream is soon identified as the Medway, and recognized for its natural supply of fish. Beyond that resource, there is the cultivated habitat of the Sidneys' fish ponds, useful especially at those times when "the high swolne Medway faile[s] thy dish." Here, an element of chance comes from the unpredictability of angling. The ponds, by contrast, are reliably teeming with piscine life and easy to plunder. There are "fat, aged carps, that runne into thy net" and "bright eeles, that ... leape on land, / Before the fisher, or into his hand."[7] The reader might extend this vision to include little Mary Sidney, fitted with rod or net, learning the art of angling by the edge of the water. More certain is that she grew up to have a neighbor with fish ponds, if not a few of her own, at Loughton. There are records of Sir Thomas Smythe digging ponds on his nearby property at Hill Hall, and of two employees of Robert Wroth visiting Penshurst to talk about pisciculture in 1612.[8] There is no reason to assume that Wroth had a love of angling, but she must have been familiar with various piscatory practices and their consequences—how they shaped the landscape, decked the table, and occasioned conversation among family members.

Fishing was also an aspect of the literary environment into which Wroth was born. Her uncle Philip Sidney's *Arcadia*, one of the major sources for Wroth's own prose romance, features many references to fish and fishing; the diversity of piscine forms and the challenge of catching them appear to have been favorite metaphors for Sidney. Of particular interest here is a scene in the first book of the revised romance (the "New Arcadia" of 1590) that figures forth an actual angling excursion—a junket led by the Amazon Zelmane, who is actually a man named Pyrocles in disguise. Having fallen in love with an image of the princess Philoclea, Pyrocles has adopted a persona that grants him access to the young woman and her family.[9] As Zelmane, Pyrocles finds his way into the service of Basilius, duke of Arcadia and father of Philoclea, but his hopes are hindered by other affections. Basilius falls in love with Zelmane, and his wife Gynecia falls in love with the man she can see through Pyrocles's disguise.[10] Daunted by the task of wooing Philoclea while besieged by the lusts of her parents, Pyrocles/Zelmane finds a way to blow off steam. She—to use Sidney's preferred pronoun for the character—makes a habit of bringing the family out on fishing trips. The narrator describes in turn how each member of the party behaves on such excursions:

> Zelmane fayled not to intice them all many times abroad, because she was desirous her friend Musidorus (neere whom of purpose

> she led them) might haue full sight of them. Sometimes angling to a little Riuer neere hand, which for the moisture it bestowed vpon rootes of some flourishing Trees, was rewarded with their shadowe. There would they sit downe, and pretie wagers be made betweene Pamela and Philoclea, which could soonest beguile silly fishes; while Zelmane protested, that the fit pray for them was hartes of Princes. She also had an angle in her hand; but the taker was so taken, that she had forgotten taking. Basilius in the meane time would be the cooke himselfe of what was so caught, and Gynecia sit still, but with no still pensifnesse.[11]

Much of what characterizes angling in the *Urania* appears in this passage, from the mutually nourishing interaction between river and trees to the utter distraction exhibited by some of the fishers. In both romances, there is a tension between the project of catching fish and the draw of other people. The characters in Sidney's scene can concentrate on either fish or love, but not both at once. This is not the solitary sport celebrated in the practical piscatory literature of the period, but an excuse to gather out of doors and be seen. Even the sisters Pamela and Philoclea, who seem to be the most serious anglers in the party, turn their sport into something more social by betting on their catch in the manner of Plutarch's Antony and Cleopatra. Their father Basilius is perhaps interested in the culinary possibilities that might emerge at the riverside, but is more likely only imagining the heat of a sexual assignation with Zelmane. His wife is similarly stirred by the spectacle of the fishing Amazon; her mind is busy with unspecified thoughts.

This chapter examines several angling episodes in the *First Part* of the *Urania*. Twice Wroth makes the sport a site of interruption, where the private contemplations of the individual fisher are broken off by the appearance of strangers who have no intention to fish. In another scene, sportfishing begins—as it does for Zelmane—as a social engagement: two lovers bring their peeved friend to the pond in hopes of cheering her up, taking it for granted that a little fishing will prove a welcome distraction. For that ill-equipped angler, the conceptual metaphors associated with the capture of fish are all too reminiscent of the wrongs she has suffered; the netted creatures are not individual lives (or even a tasty supper) but representations of her own enthrallment. Angling can provide the much-needed mental freedom that prepares its practitioners for revelatory experiences, or it can trap the fisher in a tangle of imagination that leads nowhere. The sport affords plenty of metaphorical potential for thinking through crises of desire, a bounty that can be overwhelming for the characters at the water's edge.

Gary Waller finds "recurring scenes in *Urania* and *Pamphilia to Amphilanthus* of inner spaces, private chambers and gardens, as fervently desired and protected places for women's self-contemplation without onlookers." Noting how the romance often encloses its characters within rigid social and architectural bounds, Waller contends that the outdoorsy women of the *Urania* are rare birds: "Only occasionally do women venture alone outside such confines ... there are a few women warriors, fishers, and hunters."[12] The solitary anglers of the *First Part* are interesting for their ability to find—if not maintain—private spaces out of doors. For Wroth, angling seems to belong particularly to women, and the refuge that it affords stands in sharp contrast to the expectations of sociality that interrupt or complicate the recreation. The one thing that is certain about fishing in the *Urania* is that it doesn't last long; the claims of the social world soon interfere with the wild privacy of the angler. These interruptions, it should be said, are not necessarily unwelcome; fishers seem keen to leave their rods behind and begin a new adventure. Like the distracted anglers of *Arcadia*, Wroth's fishers have to choose between meaningful social interaction and their individual worlds of private reflection.

## Girl Angler, Interrupted

As they make their way toward Hungary in book two of the *First Part* of the *Urania*, the kings Amphilanthus and Ollorandus enter a riparian landscape that is familiar to one of them. Their course runs parallel to a waterway that represents clarity and hope for Ollorandus, who automatically refers to a mental map and recognizes this river as a route to his beloved: "He welcomd each drop which passed by him, knowing, the place where his Mistris most commonly lived at, was seated on this streame" (288).[13] The river has no such meaning for Amphilanthus, whose thoughts are miles away, dwelling on his own love. "Thus they rid together," the narrator explains, "yet their thoughts so farre asunder, as might have made them strangers" (288). The entrance of the two travelers into the riparian zone is characterized by alternations of proximity and distance, familiarity and ignorance, belonging and alienation. This pattern is a reflection of the ecosystem through which Amphilanthus and Ollorandus now wend their way: a place where disparate species and elements can become entangled, as if in defiance of the differences in kind that would ordinarily prevent them from joining forces and becoming something new.

Unlike Amphilanthus, the trees that grow along the riverbank are not foreign visitors to these waters; they are visibly, intimately involved with them. The depth of their mutual familiarity is emphasized by the capital initials that are granted to both the "Osiers" and the "River" (288), a typological detail that acknowledges the unusual degree of emotionality exhibited by both parties. The description of the riverside trees here, as in Sidney's *Arcadia*, functions as a portal into a piscatory scene. A long sentence fragment carries the reader away from the visitors, into the relationship that inheres between plant and water, and toward a different human presence:

> As thus they passed among some Osiers, which grew by the side of that River, some of them within it, or the water in love with their rootes, chastly embraced them, making pretty fine ponds betweene each other, the armes, and bodyes of the trees, lying so kindly to each other, as with much ease, and fine cooleness, one might passe from one to the other.
>
> (288)

The trees, in these wet circumstances, make a public display of affection that mimics the men's proximity to one another; while the travelers "rid[e] together," the osiers lie "kindly to each other." They are willow trees, always known to thrive by waterways, and now classified as members of the genus *Salix*.[14] Their limbs are like the bendy twigs that make up Friar Lawrence's "osier cage" in *Romeo and Juliet* (2.2.7), the basket he uses to gather simples for his pharmacy.[15] On their own, the osier boughs of the *Urania* affectionately weave themselves into a resilient structure, without the interference of human hands and without any clear purpose beyond the expression of mutual love. The river joins in, pooling itself within these arboreal embraces because it is besotted with the trees. The net result is just that: an enmeshment of liquid and solid that creates some infrastructure to furnish further encounters. Although Amphilanthus and Ollorandus do not dismount and take advantage of these natural bridges, the narrator here lays a new kind of groundwork, inviting human participation in a system that is already aquatic and botanical. "One might passe" from branch to branch, and one has; we see the setting before we meet the person who inhabits it.

That person is a "dainty Maide" who has "taken up her abiding" among the osiers, using the branches as a kind of dock. In a recumbent pose that is adopted by anglers in early modern treatises but much less commonly seen today, this fisherwoman dangles her rod above the water. From this first glimpse onward, the question of whether she's actually fishing is an open one: "She had an Angle in her hand, and lay as if fishing, but her mind plac'd on a higher pleasure" (288). To "lay as if fishing" is to follow the advice of John Dennys's *Secrets of Angling*, where true piscatory humility is exemplified by the fisher's willingness to get low:

> As when a lyon coucheth for his pray
> So must he stoope or kneele vpon his knee,
> To saue his line or put the weedes away,
> Or lye along sometime if neede there be,
> For any let or chance that happen may.[16]

While the fishing rod extends the maid's reach into the water and suggests a particular intent, the narrator judges the earnestness of her endeavor by what's actually going on in the apparent angler's *mind*. The young woman's distraction is due to her own fishlike state; she has been "deceived" and "caught" by love, and the experience has left her in a poetic reverie that echoes the conceit of the narrator. "As she sate," the story goes, "she would make pretty, and neate comparisons, betweene her betraying the poore silly fish, and her owne being betrayed by the craft of love" (288). First here, and again later, as the two visitors approach her and behold her form, the maid appears piscine. In her green satin gown, aglitter with diamond buttons and open in places to reveal "her delicate skin" (289), the maid is a riparian specimen that merits a closer look. But interwoven with this shiny green apparition and the narrator's own "neate comparison," both of which build upon an inherited metaphor of the lover as a caught fish, is the rather more complicated matter of the maid's overheard musings.

The angler's mental stirring is a luxury afforded by the landscape in which we find her. Like other characters in the *Urania* and beyond, the maid positions herself in relative wilderness so that she can think out loud. In typical fashion, the resulting metaphors and music deal with her heartache and her struggle to express herself. Waller observes that "such complaints, both about love and the necessity and difficulty of putting love into words, are recorded in the poems scattered throughout the work, as well as in *Pamphilia to Amphilanthus* itself." Pamphilia, Wroth's most consistent avatar in the romance, "frequently takes refuge in the woods where she can write or be alone with her thoughts .... Such a desire for privacy articulates a deep psychological and social yearning. ... It is a desire to complete stories of her own."[17] In the case of the angling maid, that narrative completion is both deferred and facilitated by the arrival of the travelers, who draw her from her fishing hole and gradually replace her sport with the equally delicate art of conversation.

The maid's contemplations take two forms: a song and a string of spoken comparisons. It's hard to say which comes first, because the sequence of events that make up the initial encounter between the maid and the men is unclear. The narrative, in keeping with the episode's theme of distance and proximity, alternates between the two parties' perspectives and begins to describe their introduction only to double back and reestablish the gap that separates them. On the page, it is the maid's song that is first recorded verbatim. Like her rod, the music extends outward from her body and has an alluring effect; for the travelers, "the voyce [is] the cause of their seeking any body in that place" (288). Katherine Larson reminds readers of Wroth's substantial musical training, an aspect of her education that resulted in a "fascination with the affective power of song, its relationship to specific spaces of textual circulation and musical performance, and the dynamism of the singing body." In the *Urania*, she notes, women's songs are often extemporized in private and overheard.[18] There is some measure of irony, then, in the difficulty modern readers have when we try to hear the music emanating from such singers. In the absence of a printed tune (or even the title of a familiar melody), we have to remember that Wroth intended her singing characters to do more than utter lines of verse.

The lyric is addressed to a personified Love (or perhaps a deified beloved), who is invited to anatomize the singer and take her heart. It recalls the narrator's recent entrance into the maid's mind by describing the very contents of that realm: "Love peruse me, seeke, and finde / How each corner of my minde / is a twine / woven to shine" (288). In terms that reflect the maid's sport, the gleaming network of thoughts is contrasted with "a Webb ill made, foule fram'd" (288). The singer then calls for the opening up of her body, where more shining threads appear: "sinewes" and "vaines" that ultimately connect to the lover's heart (289). While there is no explicit simile linking the mental and bodily images of the song to the angler's gear, there is a distinct lexical echo that sounds across the maid's two contemplative genres. In the song, she commands, "When you thus anatomise / All my body, my heart prise;" in her spoken reflection, she says she has been hopeful "of the fishes prison." Reading retrospectively, from the noun form *prison* back to the verb *to prise*, we can see that angling has remained on the lover's mind.[19] The strong grasp of love is consistently, if subtly, associated with the angler's power to take a fish, from the narrator's initial description of the maid ("the hooke of love having caught her so fast, as nothing could release her" [288]) to the images of capture in both the song and the monologue.

In strictly textual terms, Amphilanthus and Ollorandus first approach after the maid's song, and appear to receive an immediate welcome: "The song ended, they went towards her … she curteous and excellently witty, gave them entertainement" (289). The maid is then described in her resplendent attire, then (again) with her rod in hand, talking to herself; in this second iteration, it is her private speech rather than her song that gets interrupted. She stands up to greet them, abandoning the horizontal pose of the angler and the project of quiet reflection in order to show the princes due hospitality. This sudden and easy shift of attention would probably be disconcerting to early modern proponents of sportfishing, who often describe their hobby as a welcome departure from social life. In William Samuel's 1577 treatise, the character Piscator expresses fears that non-anglers might come along and spoil his recreation—he tells his companion Viator to keep his voice down because "I would not gladly by your loud talking that either some bungler, idle person, or jester might thereby resort unto us." Hell, for this angler, is other people.[20] Both John Dennys and the author of the medieval *Treatyse* take it for granted that the sportfisher derives pleasure and spiritual health from being alone. In the context of the available models for piscatory success, then, Wroth's fishing maid seems unusually friendly.

The doubled account of the travelers' approach makes no literal sense, but participates in a pattern of repetitions and echoes that is the organizing principle of Wroth's romance. For Gary Waller, "the stories in *Urania* are at once all different and yet the same ... like the narratives of the unconscious or compulsive daydreams."[21] In Rahel Orgis's treatment of the romance's "interlace structure," the sometimes bewildering narrative doublings are all part of how the text works to shape its readers.[22] Here at the riverside, the narrator provides several versions of events, all of which contribute to the vaguely supernatural atmosphere of the broader episode. The reader's vision is appropriately blurred by the maid's caprice in turning abruptly away from her sport, the ambiguity surrounding the order of her speech and song, and the rapid shifts between the host's perspective and that of her guests. This is a world of sudden changes.

The confused, dreamlike chronology of the encounter matches the flickering quality of the maid's mental experimentation. Unlike the tidy metaphor of lover-as-fish, the figure subsequently presented by both the narrator and the maid is anything but "neate": It is a conceit gone wild, an addled riddle that finds no stable position for its various parts and is ultimately left by the wayside with the maid's neglected rod. The narrator introduces the monologue with a simile that cannot sustain itself through the rapid semantic shifts offered up by the maid herself:

> She held her angle as neglectively, as love the ill causer of her paine held her, when the poore little fish did plaie with the baite, or offer to swallow it, too big for them, yet made the cork stirre: "So," would she say "doth Love with me, play with me, shew me pleasures, but lets me enjoy nothing but the touch of them, and the smart of the hooke that hurts me without gaine, and only gives as light a good to me, as the hope this floting corke did give me of the fishes prison."
> (289)

The ideas that love is a cruel angler and that its seeming pleasures only smart are compatible with the conceit that has already been established. But new and complicating notions enter when the lover herself is suddenly represented as a fishing rod. Now the maid is "held neglectively" by love, dangled idly in a position from which she can see what she wants but not secure it. She has become an ineffectual tool.

The maid's elaborate analogy, which begins "So ... doth Love with me," has no clear referent. It may mean—in a continuation of the narrator's simile—that love is to the forlorn lover as an angler is to her rod. It may also mean that the fish who tease her by merely nudging the bait have glided in and supplanted the love-as-angler metaphor. In the latter case, the little fish frustrate the angler through a combination of highly visible feeding behavior and an ultimate failure to bite. She is shown much but can "enjoy nothing" beyond the sensation of movement at the end of her line and the painful insult brought on by faulty tackle. The hook that is designed to capture fish cannot take its intended effect; instead, it doubles back to injure its owner with acute disappointment. In turn, the hook is compared to the deceptively wobbling cork on the water's surface, whose lightness is both physical and figurative—when it stirs, it delivers an airy, weightless promise that someone below has taken the bait. A good cork, of course, should not send false signals. In *The Secrets of Angling*, each cork must be trimmed with a razor so that one end is bigger than the other. Such shaping makes the cork a better indicator: "The smaller end doth serue to sinke more light / .... The greater swims aloft and stands vpright;" this way, the line and the bait are kept "at euen stay, / That when the Fish begins to nib and byte, / The mouing of the float doth them bewray."[23]

Although the maid's song has declared that faith and love bring both "gaine and sting," her piscatory monologue shows that it is possible to receive pain "without gaine," to have one's hopes dashed utterly. This disillusioned view of love comes with the speaker's shift from thinking of herself as a fish to reflecting (perhaps more earnestly) on her current experience of fishing with the wrong gear. The simpler comparison—of the lover's hunger to the appetite of the fish—gives way to an image of the disappointed angler who can *prise* no catch from the river. Soon the maid fully discards her recreational tools and replaces private musings with social interaction. As the two princes approach, her "imaginations [are] laid aside" (289). The use of the plural "imaginations" acknowledges that the foregoing conceit is slippery and dynamic at best, if not hopelessly elusive. In the single suite of comparisons, the maid is a fish who suffers at the cruel hand of the angler, an angler betrayed by the gear that had seemed to guarantee success, and a neglected fishing rod fitted only with overlarge bait and a tiddly cork.

In the second account of the meeting between the angler and the travelers, the maid's gestures emphasize her altered orientation, from the prone horizontality of introspection to a vertical pose of greeting, as the narrator pairs the motion of ejecting the rod with the decorous signal of welcome made possible by the freed hand: "she presently rose up, and with fit and excellent respect she saluted them, throwing downe her rod, wherewith she had thretned, but executed little hurt" (289). The dysfunctional relationship with the angle ends (temporarily, at least) with the imperative of human hospitality. Having unwittingly lured these foreign creatures into her company, the maid accepts her duty as host and conveys them to a setting that is characterized by sociality rather than solitude. The host brings her princely guests away from the botanical wilderness of her fishing hole, drawing them by degrees to an enclosed human habitation:

> Passing over her transparent bridge, [she] went unto them, whom she used according to their merits, when shee understood who they were, conducting them along that river to a most curious garden, and so into another, and then into as rare a place for building, as they in all their travels had ever seene, being a house of white Marble. Varietie of all delights were there, and shee desirous that they should enjoy them, giving all the free and noble welcome that could be afforded.
>
> (289–90)

The porous, plant-made structure of the osier "bridge" is abandoned for the spectacle of two horticultural spaces, further along the river, that are characterized by their opaque limits. Each of the gardens is an enclosure with an entrance and an exit, in contrast to the exposed arboreal network in which the angler has been spied. With their combination of botanical and human agency, the two gardens prepare the visitors for the more distinctly human-made, solid surfaces of the marble house. There, the activities of the riverbank are replaced by the "delights" of the house; the angler's solitary "kind of bed" made of intertwined osier boughs is left behind for "two marvellous faire beds" where the guests can rest, according to their stated preference, side by side.

If the openness of the riparian scene affords Amphilanthus and Ollorandus a view of the maid's private musings, the splendid interior of her home gives them a glimpse of her awkward marriage. Despite the kind hospitality of the lord of the estate, who is keen on "shewing them al the rare delights he had about his house," the visitors remain focused on the lady (as she is now called) and recognize that her feelings for her husband are "distant … from affection" (290). Amphilanthus quickly concludes that she is "one of their fellow prisoners in the Castle of love" and contrives to hear her story (290). This undertaking requires a return to the outdoors, specifically to a landscape dominated by plant life and running water:

> They walked along a dainty pleasant brooke, the bankes whereof were of such sweetnesse, as the plenty of the most delicate smelling flowers could yeeld unto them, being shadowed from the Sunnes parching by Mirtle, and that Mirtle preserved by high, and brave trees.
>
> (290)

Baiting his new friend, Amphilanthus knowingly declares that this perfect contemplative spot is "fitter much for lovers, than for such free Princes as live here" (290); in order to show just how enfettered she is, the lady promises to trade stories with him, and we thus receive the long version of her earlier lamentation.

With the acquaintances restored to the company of sustaining trees and water, this episode invites comparisons between the lady's private sport and the discursive recreation of storytelling. As a solitary angler, she has compressed her experiences into song and conceit; as an interlocutor, she delivers a detailed account of the central struggle of her life.[24] Rod and line are notably missing from the second outdoor scene, as if angling has been replaced by a conversational sport that seeks to capture stories rather than fish. In structural terms, too, the inset tale is like a creature drawn out of its habitat, brought up to the surface and exposed to view. And because it is a narrative that makes its teller newly, wholly visible to her audience, it recalls the lady's first appearance as a riparian specimen that must be approached with caution. At last, thanks to Amphilanthus's careful angling, we get to hear the substance of her song and see inside her "delicate skin."

## Separate Sport, Shared Revelation

The idea of a revelatory chance encounter that begins at the site of angling is repeated later in the *First Part*, when the Achaian queen Dalinea and her husband Parselius take up separate sports in a remote redoubt. They set up camp "hard by a wood, a delicate brooke running before them, over which was only a ford but no bridge; there the Queene fished, while the King hunted" (518). Unlike the unnamed maid of the earlier episode, Dalinea has no time either to catch fish or to wax poetic, because the narrative immediately places her in the purview of another character: "As she was at her sport, an ancient fine woman appeared on the other side smiling, and calling her by signes unto her" (518). The narration lets the reader see this mysterious person before the queen herself does, so we watch the oblivious angler—from the perspective of the other woman—and wait for her eye to be caught. But even as the two women are distinguished from one another by their different motivations for being by the river and by the physical barrier of the river itself, they become equivalent in their active pursuit of other bodies. The mirroring effect is emphasized by the older woman's movement in parallel with Dalinea, and by the apparent ease with which Dalinea interprets her gestural language:

> The Queene was at first carelesse of her, but she still following right against her, calling as she went with her Angle fit for the streame, shee at last marked her, and was so intised [enticed] as she resolved to goe with her …
>
> (518)

The ambiguous use of pronouns in this sequence only adds to the uncanny impression that Dalinea is being angled to the far shore by a version of herself.

In keeping with the pattern established earlier in the romance, the angler quickly and decisively abandons her sport and crosses over into the company of a stranger. The deliberate pose of the sportfisher, "with her Angle fit for the streame," is instantly undone by a sudden, overwhelming desire to follow the old woman. Piscatory gestures are replaced with the motions of command: Dalinea "send[s] straight for her Horse" and some of her servants, instructing one of them to convey the old woman as they all ride off in the direction specified by the stranger (518). The company passes through several apparently enchanted habitats, including "a place thicke, and scratching" where the thorny shrubs are keen to grab and hold their visitor, and "a little round place" reminiscent of a fairy circle. When the stranger tells her to dismount, Dalinea "obey[s] as if led by Fortune," and the two, now magically left alone, make their way by turns into a lavish hall with an altar at one end. Here, Dalinea is unexpectedly reunited with her husband, who has similarly been drawn away from his hunt by an older stranger. The altar is revealed to be the "Tombe of Love;" it holds a pair of half-siblings who unwittingly fell in love, died "at one instant" (525) without committing any physical act of incest, and were enshrined by their grieving father Distantes (great-uncle to Dalinea). The tomb and its staff have been waiting for the day "when a woman should raigne heere, loving as well, as much deserving, but more happily enjoying to her comfort, which is your selfe" (526).

Dalinea's joyful love for Parselius is explicitly contrasted with the overwhelming shame and misery that Distantes's unnamed daughter experiences upon learning that her beloved is her brother. The king and queen express their pity for the star-crossed lovers by endowing the tomb with a "large allowance ... to maintain lights for ever over them" and promising that their family will see to "the keeping of the place" (526). But the revelation also brings a distinct pleasure: "made happy with the rarest accident of love, and the richest Tombe Love in the world had, they return'd with all content ... to their setled Court" (526). Dalinea and Parselius have come into the wilderness to pursue two different kinds of prey in contiguous habitats; instead, they are themselves carefully apprehended according to a long-standing plan, brought to a single, particular location, and told the story of two bodies that were trapped long ago by an illicit love. The queen and her consort go straight home, presumably satisfied with the events that have interrupted and replaced their sports. Just as quickly, the narrative leaves them behind and follows other characters.

Through the divergent recreations of Dalinea and Parselius, the typical early modern distinction between fishing and other predatory sports is maintained. The king and queen are engaged in parallel pursuits, but they must part ways in order to take up their projects in earnest. Their necessary separation into different habitats is itself charged with affective potential, making the surprise of their sudden reunion all the more striking:

"He beheld her, and she him, but as two Statues, set with their eyes one upon the other without power to speak: so stood they, admiration filling them as they were but wonder" (519). The theme of distance and proximity is at the heart of the tragedy enshrined in the Tomb of Love—the young lovers, too close in blood, can only finally be parted by death. Their father's name echoes the distance that they have taken pains to maintain from one another, at first limiting themselves to what "chast[e] love did warrant" (520) and later "part[ing] their misery" (524) to preserve their family's honor from the taint of incest. In the early stages of their story, they speak as lovers often do of the pain of being apart, not knowing that they are bound in a tragic proximity. Dalinea and Parselius, conversely, can blithely bid each other adieu for an afternoon. Hearing the story of her doomed cousins, the new monarch of Achaia is encouraged to see herself as one "more happily enjoying to her comfort" because her mind is at ease in both the presence and the absence of her beloved.

Although the episode only begins to tell of actual angling, Dalinea's solitary fishing excursion sets her in a confident balance with her hunting husband and acts as an emblem of her good fortune in love. And while the sport does not lead directly to the revelation at the Tomb of Love, we can recognize this sequence as the second in which an angler's private scene of recreation unfolds to become the starting point for a new adventure. The nameless fisherwoman of Book Two conveys Amphilanthus and Ollorandus through several distinct outdoor spaces before they reach a splendid building; similarly, the mysterious old woman of *Book Four* brings Dalinea into brambles, across clearings, and down a hidden stairwell to the underground tomb. In both cases, the postures of angling are interrupted by the approach of a stranger, and inviting gestures are flung across a distance to initiate a human encounter that will include a lengthy narration on the subject of love. Whether love has left them happy or sad, Wroth's anglers are quick to leave the conventionally contemplative topos of the riverbank and start something new.

## Not Fun for Everyone

A third angling episode from the *First Part* of the *Urania* shows the sport being administered from outside as a kind of psychic medicine. Here, too, the fisher's patience is limited. Amphilanthus, the king of Naples, finds himself entertaining two admirers near the end of book two: his cousin and lover Pamphilia, and the equally enamored Antissia. The visit is awkward from the start; Pamphilia and Amphilanthus greet Antissia with a gesture that links all three lovers together and fills the visitor with unwarranted hope:

> Antissia was so much joy'd as she was but that cosening thing it selfe, ravished with false delight; she triumphed in the blaze, while the true

> fire burnt more solidly, and in another place. She was conducted to the Palace, Pamphilia with her lefte arme embracing her, holding Amphilanthus with the right hand.
>
> (319)

The king welcomes Antissia's affections—"for what man lives, that glories not in multitudes of womens loves?" (325)—but does not satisfy her ardor. The guest, suffering from immoderate happiness in the presence of Amphilanthus, cannot keep her mouth shut. While she prattles on, the two cousins exchange knowing glances about Antissia's "over-esteemd good fortune" (320) and generally endeavor to show her a good time.

When the three characters sit down for a typical sylvan storytelling session, only Antissia's contribution is related by the narrator. As she speaks of her encounter with a "Brittaine Lady" who once told her of her misfortune in love, the teller gradually abandons the third person and begins to perform the role of the spurned lover. In Tiffany Werth's reading, this shift "reveals of course that the story she tells is her own story, not that of some 'Lady.' " The confusion evident in Antissia's slippery pronouns accords, for Werth, with a broader "interpretive problem … stem[ming] from her refusal to maintain boundaries between fantasy and reality," a pattern that will shape this character's experience deep into the *Second Part* of the romance.[25] In her story, Antissia's lady begins to suspect her beloved of getting "an other love;" he pays her an expected visit, but leaves her chamber after only "a little talke" and a courteous kiss (324). If Antissia is a shoddy storyteller, barely able to conceal the true identity of her protagonist, Pamphilia and Amphilanthus make an equally bad audience. They beg her to continue, but seem to snigger: "Pamphilia smild to heare her come to that; the King was forc'd to cover his conceits, and wish her to proceed" (324). Already having revealed too much, Antissia spurts out "Pray God … I doe not play the Brittaine Lady now" and quickly wraps the story up "with greater passion" (324).

Recognizing that the exchange of stories has not resulted in mutual satisfaction, the lovers extend their hospitality in another direction:

> They found she was not pleas'd, therefore they sought some other way to please, and rising walked into another wood, and so unto a pond, which they did fish, and passe the time with all, while poore Antissia thought her selfe each fish, and Amphilanthus stil the nette that caught her, in all shapes, or fashions she could be framed in.
>
> (324)

Here, angling is a quiet alternative to the potentially volatile art of conversation. It brings its practitioners together in a single location, but allows them to inhabit distinct spaces of individual reflection. While this recreational shift releases Pamphilia and Amphilanthus from their obligation to listen and respond with sensitivity, it does little to ease Antissia's jealous mind. Come evening, Amphilanthus is once again trapping her with his deceptive attentions and seeming to be enthralled by her glances: "He, though now neither fond nor loving to her, yet seem'd to like her love, if only that his might be the more prized, wonne from so brave and passionate a lady; and thus she often caught his eyes" (325).

In the hastily drawn scene by the edge of the pond, the narrator's language echoes that of Shakespeare's Cleopatra, but with the roles reversed. Antissia thinks "her selfe each fish;" the Egyptian queen projects that she will "think them every one an Antony" (2.5.14). In both cases, recreational fishing is presented as a distraction from the pain of envy—more specifically, from the suspicion that one may have been caught and released by a fickle lover. For Antissia, the plan backfires. Because of the ambiguous use of *they*, it is not clear whether the sullen visitor even participates in the sport in the first place; if she does, she may lack the skill that would make the activity a balm to her wounded pride. Like the unnamed lady who dangles a rod earlier in book two, this potential angler is preoccupied by her "imaginations," unable to read the sport as anything other than a metaphor for her own vulnerability and disappointment.

Distinguishing this brief episode from the other scenes of angling in the *Urania* is the absence of references to rods and lines; the salient piece of equipment is instead the net, a tool that would have been indispensable to sportfishers before the widespread use of reels. The image of a lover "framed" by a net recalls the forlorn angler's song—the one overheard by Amphilanthus and Ollorandus—in which the lover describes her thoughts as "a twine / woven to shine" rather than "a Webb ill made, foule fram'd" (288). The verb *to frame* applies to both the construction and the function of the net;[26] it is a woven container that has been designed to reveal its contents. Feeling "framed," Antissia silently conjures an image that multiplies her suffering. Even if she doesn't speak aloud, Antissia joins the company of the many women in Wroth's works who, in Naomi Miller's terms, "appear as speaking subjects with an emerging capacity, if not always sufficient assurance, to fashion selves through discourse." Reading across Wroth's genres of poetry, drama, and prose, Miller adopts Luce Irigaray's formulation "neither one nor two" to describe the redefinition of women's selfhood in the *Urania* and beyond:

> Transfiguring the fragmented objectification of female sexuality which characterizes the culturally determined role of dark lady in the conduct manuals and marriage treatises, sonnet sequences and romances of the period, Wroth represents a variety of female characters whose voices come to define feminine parameters for a figuration of subjectivity that is "neither one nor two," neither singular nor dual, but multiple.[27]

While the mechanisms of rod, line, hook, and bait might serve to represent a contest between two parties, the net metaphor befits the spurned lover's experience of a complex subjectivity that just can't win. All of the many possible Antissias, no matter how they express themselves, are subject to the same cruel fate at the capturing hand of Amphilanthus. Past, present, and future selves, regardless of "shape" or "fashion," and despite the lover's apparent ability to change, are equally, inevitably besotted. Because the net, unlike the single baited line, can hold several fish or even several species at once, it provides a more poignant point of identification for a character who knows her own "infinite variety" (to borrow from *Antony and Cleopatra*).

Although Pamphilia and Amphilanthus choose fishing as "some other way to please," a surefire pastime that is assumed to have wide appeal, the sport's tendency to remind its practitioners of other things makes it much less fun for their forlorn visitor. Just as the unnamed lady angler of the earlier episode finds herself composing elaborate conceits rather than catching any fish, Antissia's experience of the sport is limited to a private exercise in poetic introspection. Framing herself as a victim of love, and thinking with eloquence about how she might interpret her position, she demands some share of readerly sympathy. The narrative voice leaves the jocund couple in the distance to relate Antissia's pondside reflections. Pamphilia and Amphilanthus's carefree mutual recreation is exceptional in the world of the *Urania*, where angling is overwhelmingly solitary. Like Antony and Cleopatra's fishing, this couple's sport is not without negative consequences for other people. Antissia is much more typical; she resembles the distracted, unsuccessful anglers of Sidney's *Arcadia*, and, with her sense of a dynamic, multiple selfhood and her frustrated desire to express herself in language, she is a quintessential Wrothian angler.

## Notes

1 Margaret P. Hannay, *Mary Sidney, Lady Wroth* (Farnham, Uk: Ashgate, 2010), 118.
2 L. L. Ford, "Wroth, Sir Robert (c. 1539–1606), landowner and politician," *DNB* (September 2004), https://doi.org/10.1093/ref:odnb/30083.
3 Mary Ellen Lamb, "Wroth [née Sidney], Lady Mary (1587?–1651/1653), author," *DNB* (January 2008), https://doi.org/10.1093/ref:odnb/30082.
4 Lamb, "Wroth [née Sidney], Lady Mary."
5 Ben Jonson, "CV. To Mary Lady Wroth," in *The Workes of Benjamin Jonson* (1616, STC [2nd ed.] 14751), 801–802.
6 Hannay, *Mary Sidney, Lady Wroth*, 176, 162, 154.
7 Ben Jonson, "To Penshurst," in *The Workes*, 820.
8 Hannay, *Mary Sidney, Lady Wroth*, 144, 176.
9 Philip Sidney, *The Countesse of Pembrokes Arcadia* [The New Arcadia] (1593, STC [2nd ed.] 22540), 25v-26r. The first edition of the revised *Arcadia* was printed in 1590.

10 Sidney, *Arcadia*, 26r-27r, 28v.
11 Sidney, *Arcadia*, 29r.
12 Gary Waller, *The Sidney Family Romance: Mary Wroth, William Herbert, and the Early Modern Construction of Gender* (Detroit, MI: Wayne State University Press, 1993), 103, 272.
13 Lady Mary Wroth, *The First Part of the Countess of Montgomery's Urania*, ed. Josephine A. Roberts (Binghamton, NY: Medieval and Renaissance Texts & Studies, 1995). All text citations refer to page numbers in this edition.
14 *OED*, s.v. "osier, n. and adj.," accessed April 20, 2018, www.oed.com/view/Entry/132987;jsessionid=7C59B959B782DD099C822B2405040E44?redirectedFrom=osier.
15 William Shakespeare, *Romeo and Juliet*, ed. Francis X. Connor, in *The New Oxford Shakespeare: Modern Critical Edition*, eds. Gary Taylor et al., online edition (Oxford: Oxford University Press, 2016), DOI: 10.1093/actrade/9780199591152.book.1.
16 [John Dennys], *The Secrets of Angling* (1613, STC [2nd ed.] 6611), D6r.
17 Waller, *The Sidney Family Romance*, 280, 279.
18 Katherine R. Larson, "Voicing Lyric: The Songs of Mary Wroth," in *Re-Reading Mary Wroth*, ed. Katherine R. Larson and Naomi J. Miller (New York: Palgrave Macmillan, 2015), 119, 129.
19 Interestingly, the verb *to prise* and the noun *prison* come from different origins (*OED*, s.vv. "prison, n." and "prise, v.," accessed April 20, 2018, www.oed.com/view/Entry/151652?rskey=BDe63O&result=6&isAdvanced=false and www.oed.com/view/Entry/151652).
20 [William Samuel], *The Arte of Angling* (1577), ed. Gerald Eades Bentley (Princeton, NJ: Princeton University Press, 1956), 17.
21 Waller, *The Sidney Family Romance*, 250.
22 Rahel Orgis, *Narrative Structure and Reader Formation in Lady Mary Wroth's* Urania (Abingdon, UK: Routledge, 2017), 190, 1.
23 Dennys, *The Secrets of Angling*, B3r.
24 Readers have advanced several hypotheses about the autobiographical dimensions of the lady's tale. Josephine Roberts identifies the lady's undesirable husband as a reflection of Robert Wroth; Charimellus is "a socially inept, bumbling man, pitied by the lady more than admired" (introduction to *The First Part of the Countess of Montgomery's Urania*, xc). Margaret Hannay speculates that the lady's trouble in love may allude to the marriage negotiations of Wroth's cousin, Lady Anne Herbert, and the elderly Earl of Hertford (*Mary Sidney, Lady Wroth*, 91–93). She also proposes that Wroth was thinking of her own awkward fit among her Puritan in-laws when she composed "the quasi-autobiographical tale of the angler lady" (120). Jennifer Lee Carrell numbers the story among the romance's many iterations of a "ghostly ur-tale" in which a woman falls in love with a childhood friend and remains devoted to him despite his faithlessness ("A Pack of Lies in a Looking Glass: Lady Mary Wroth's *Urania* and the Magic Mirror of Romance," *Studies in English Literature 1500-1900* 34 [1994]: 94). Such stories are, among other things, reflections of Wroth's own relationship with her cousin William Herbert, Earl of Pembroke. For the purposes of my analysis, the content of the inset tale is less important than how we come to hear the story in the first place.

25 Tiffany Jo Werth, *The Fabulous Dark Cloister: Romance in England after the Reformation* (Baltimore, MD: Johns Hopkins University Press, 2011), 138.

26 *OED*, s.v. "frame, v.," accessed April 20, 2018, www.oed.com/view/Entry/74152?rskey=0eNU4H&result=3&isAdvanced=false.

27 Naomi J. Miller, *Changing the Subject: Mary Wroth and Figurations of Gender in Early Modern England* (Lexington: University Press of Kentucky, 1996), 62–63.

## References


Carrell, Jennifer Lee. "A Pack of Lies in a Looking Glass: Lady Mary Wroth's *Urania* and the Magic Mirror of Romance." *Studies in English Literature 1500–1900* 34, no. 1 (Winter 1994): 79–107.

[Dennys, John]. *The Secrets of Angling*. London, 1613. STC (2nd ed.) 6611.

Ford, L. L. "Wroth, Sir Robert (c. 1539–1606), landowner and politician." In *Oxford Dictionary of National Biography*. Oxford University Press, September 2004. DOI: https://doi.org/10.1093/ref:odnb/30083.

Hannay, Margaret P. *Mary Sidney, Lady Wroth*. Farnham, UK: Ashgate, 2010.

Jonson, Ben. *The Workes of Benjamin Jonson*. London, 1616. STC (2nd ed.) 14751.

Lamb, Mary Ellen. "Wroth [née Sidney], Lady Mary (1587?–1651/1653), author." In *Oxford Dictionary of National Biography*. Oxford University Press, January 2008. DOI: https://doi.org/10.1093/ref:odnb/30082.

Larson, Katherine R. "Voicing Lyric: The Songs of Mary Wroth." In *Re-Reading Mary Wroth*, edited by Katherine R. Larson and Naomi J. Miller, 119–36. New York: Palgrave Macmillan, 2015.

Miller, Naomi J. *Changing the Subject: Mary Wroth and Figurations of Gender in Early Modern England*. Lexington: University Press of Kentucky, 1996.

Orgis, Rahel. *Narrative Structure and Reader Formation in Lady Mary Wroth's Urania*. Abingdon, UK: Routledge, 2017.

Roberts, Josephine. Introduction to *The First Part of the Countess of Montgomery's Urania*, by Lady Mary Wroth. Binghamton, NY: Medieval and Renaissance Texts & Studies, 1995.

[Samuel, William]. *The Arte of Angling*. 1577. Edited by Gerald Eades Bentley. Princeton, NJ: Princeton University Press, 1956.

Shakespeare, William. *Romeo and Juliet*. In *The New Oxford Shakespeare: Modern Critical Edition*. Edited by Gary Taylor, John Jowett, Terri Bourus, and Gabriel Egan. Online edition. Oxford: Oxford University Press, 2016. DOI: 10.1093/actrade/9780199591152.book.1.

Sidney, Philip. *The Countesse of Pembrokes Arcadia* [The New Arcadia]. London, 1593, STC (2nd ed.) 22540.

Waller, Gary. *The Sidney Family Romance: Mary Wroth, William Herbert, and the Early Modern Construction of Gender*. Detroit, MI: Wayne State University Press, 1993.

Werth, Tiffany Jo. *The Fabulous Dark Cloister: Romance in England after the Reformation*. Baltimore, MD: Johns Hopkins University Press, 2011.

Wroth, Mary. *The First Part of the Countess of Montgomery's Urania*. Edited by Josephine A. Roberts. Binghamton, NY: Medieval and Renaissance Texts & Studies, 1995.

# 6 The Quickening of Walton's Incomplete Angler

Izaak Walton's monumental treatise *The Compleat Angler*, first published in 1653, served as both a retrospective of received piscatory knowledge and a sudden promotion of the sport in the cultural imagination of early modern England. Capacious enough to accommodate material from a whole library of fishing manuals, natural histories, and poetry collections, but small enough to be carried with ease to the riverbank, the *Angler* would become one of the most enduring and celebrated books in English literary history. And although it is in some respects a very bookish book, with its ample citations and its appeal to "the contemplative man," the fundamental structure of the text is dramatic: it is the many-voiced story of a burgeoning friendship and one man's conversion to the "Brotherhood of the Angle." Within the already flamboyant dialogue, the expert angler Piscator's teaching and proselytizing are often interrupted, many times by the speaker himself, in favor of some other genre of verbal expression: anecdote, recipe, verse recitation, or song. In addition to these generic shifts are many non-verbal interruptions consisting of images. At such turns, the function of the little book itself can suddenly change.

Perhaps the most striking example of this dynamism is the very first piscine illustration in the *Angler*: a trout that appears right on cue. The angler Peter shares his plans to catch a fish for his friend Coridon, and Piscator immediately presents him with the desired bounty:

PETER. Brother Piscator, my friend is an honest Country man, and his name is Coridon, a most downright witty merry companion that met me here purposely to eat a Trout and be pleasant, and I have not yet wet my line since I came from home: But I wil fit him to morrow with a *Trout* for his breakfast, if the weather be any thing like.

PISC. Nay brother, you shall not delay him so long, for look you here is a Trout will fill six reasonable bellies. Come Hostis, dress it presently, and get us what other meat the house will afford, and give us some good Ale, and lets be merrie.

(80–81)[1]

The trout is hauled up into the textual framework of the unfolding narrative, as Piscator points it out to both his reader and his interlocutor: "Look you here." In this instance, the image of the fish is revelatory and funny, because it mixes the reader's actual position with that of the character Peter, thus inviting any of us into the Brotherhood of the Angle. The piscine illustrations in the *Angler*, most of which are based on engravings from Gessner's *Historiae animalium*,[2] are often interposed between information about the species' life histories and instructions on how to catch or even cook them; thus, they allow the animals to be seen several ways at once. Similarly, the student of angling is positioned simultaneously at multiple locations: out at the riverside, indoors with the book, in the kitchen, and at table.

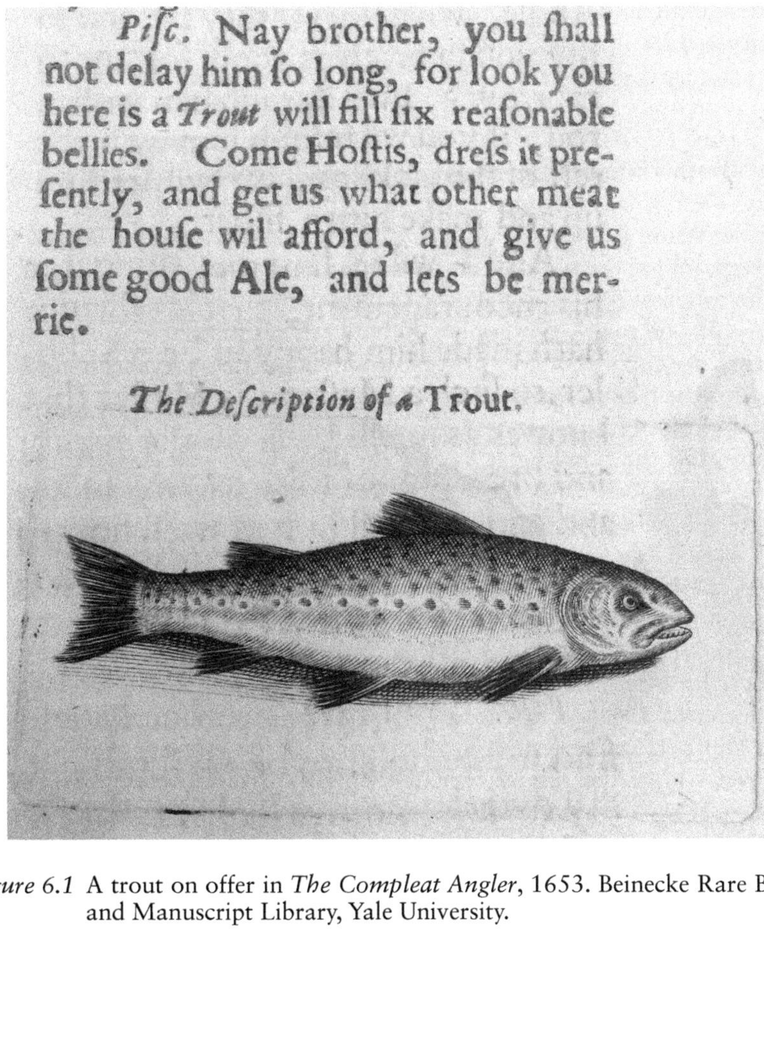

*Piſc.* Nay brother, you ſhall not delay him ſo long, for look you here is a *Trout* will fill ſix reaſonable bellies. Come Hoſtis, dreſs it preſently, and get us what other meat the houſe wil afford, and give us ſome good Ale, and lets be merrie.

*The Deſcription of a* Trout.

*Figure 6.1* A trout on offer in *The Compleat Angler*, 1653. Beinecke Rare Book and Manuscript Library, Yale University.

As the *Angler* shifts its function from cookbook to field guide to dramatic dialogue, it repeatedly reinterprets the reader's role. The text's composite nature, which has been most thoughtfully documented by Marjorie Swann,[3] is exemplified by its inclusion of a tune, "The Anglers Song," laid out in staves of music on two facing pages. In the first edition of the book, the two singing parts, marked "cantus" and "bassus," were set in opposite orientations so that both registers could be sung at once, with tenor and bass holding the book between them and facing one another (216–17). Here, the *Angler* becomes a performance text, one that can apparently draw people around itself in order to bring about a specific social experience and an aesthetic product. The song is introduced by Viator, who asks his angling teacher Piscator to join in: "Come, Master, you can sing well, you must sing a part of it as it is in this paper" (215). Piscator later invites the whole assembly to participate: "Come, we will all joine together, mine Hoste and all, and sing my Scholers Ketch [tune] over again" (218). The solitary reader is brought into an imaginary company of fisher-singers who promise, through their shared music, to "banish all sorrow, and sing till to morrow, / And Angle, and Angle again" (216). The inclusion of the actual tune means that readers can either envision a place for themselves in the Brotherhood of the Angle, or physically embody that new role by raising their voices in song. Either way, this is a late flourish in the vivid documentation of Viator's initiation, setting an example for others who may seek an experience of conversion and communion.

The *Angler* is conversional in both its content and its function; it relates a story of one man's fundamental change from hunter to angler, and it attempts to carry readers through a similar process by bringing them along for the fishing trip that serves as the instructional frame. The text is tidily organized into episodes, beginning with the first encounter between Piscator and the non-angler Viator. (In subsequent editions, the traveler Viator is replaced with the hunter Venator and the falconer Auceps.) There are long verbal digressions, but the relationship between "master" and "scholar" continues to develop until their bittersweet parting at the end of the book, when they renew their promise to meet and shop for tackle on the ninth of May (229, 246).

Viator's conversion is dependent upon his growing affection for his teacher; he becomes devoted to angling through his love for Piscator, which is itself fostered by the landscape and climate surrounding the two men. This persuasive system is consistent with the shape of Walton's project as a whole; the *Angler* repeatedly shows, from its prefatory matter onward, that love among human beings is the foundation for the sport's promotion and adoption. The appeal of sportfishing is, throughout Walton's book, an affective phenomenon that develops naturally from one man's love and admiration for another. Viator's conversion is doubly situated, in both the material environment of Walton's text, with its frequent depictions of great men who are also great anglers, and the imagined environment of the riverbank, endued by the classical tradition with lively affective and erotic potential. The reading that follows traces that conversion as it is narrated in the first edition of the *Angler*, with occasional comparisons between the 1653 text and the revisions published in the subsequent decades.

## Taking It Outside

The *Angler* opens, as so many early modern texts do, with a group of men gathered in a scene of mutual admiration. The prefatory material includes, first, Walton's dedication to John Offley: a letter that identifies the addressee as a "Master" of angling and the author as "Your most affectionate Friend" (A3v, A4v). Walton claims that those with less piscatorial experience would be inspired by seeing Offley in pursuit of fish: "If common Anglers should attend you, and be eye-witnesses of the success, not of your fortune, but your skill, it would doubtless beget in them an emulation to be like you" (A3r). At the outset, it is the observation of another man, rather than the weather, the gear, or the fish themselves, that might stir a person to become a better angler. Thus Walton emphasizes the proximity of student to teacher—a close enough range that the learner can be an "eye-witness" to the techniques and triumphs of the master. Starting here, and echoing himself throughout the *Angler* in the voice of Piscator, Walton cautions readers that the textual presentation of information offered in his book can never take the place of actual experience.

In the first place, the very idea for the book is something that Walton has inherited from a friend:

> I remember Sir Henry Wotton (a dear lover of this Art) has told me, that his intentions were to write a discourse of the Art, and in the praise of Angling, and doubtless he had done so, if death had not prevented him; the remembrance of which hath often made me sorry; for, if he had lived to do it, then the unlearned Angler (of which I am one) had seen some Treatise of this Art worthy his perusal, which (though some have undertaken it) I could never yet see in English.
> (A3v-A4r)

Although he is attempting to fill a perceived gap in the common store of knowledge by setting down a fishing treatise in English, the author openly admits to being a neophyte—an "unlearned Angler" whose understanding of the sport is limited. It's possible, tells Offley, that the present volume "can contribute nothing to your Knowledge" (A4v). Walton nevertheless takes up the challenge of following in Wotton's footsteps, knowing all the while that his lack of expertise may hamper the success of his treatise. The question of the text's worthiness to be read is so central for

Walton that it appears on the frontispiece in the form of litotes: the book advertises itself as "Not unworthy the perusal of most Anglers." In addition to the author's anxiety—or strategic show of humility—in regard to publishing such a volume, there is the further problem of how much can be learned about angling through reading alone.

In the epistle to the reader, which follows the dedication to Offley, Walton compares himself to the author of a fencing manual, recounting that this Mr. Hales "was laught at for his labour" because "that Art was not to be taught by words; nor is the Art of Angling" (A6v). In the revision of this letter that appeared two years later, the excuse is extended through a subsequent comparison of sportfishing to mathematics, "that

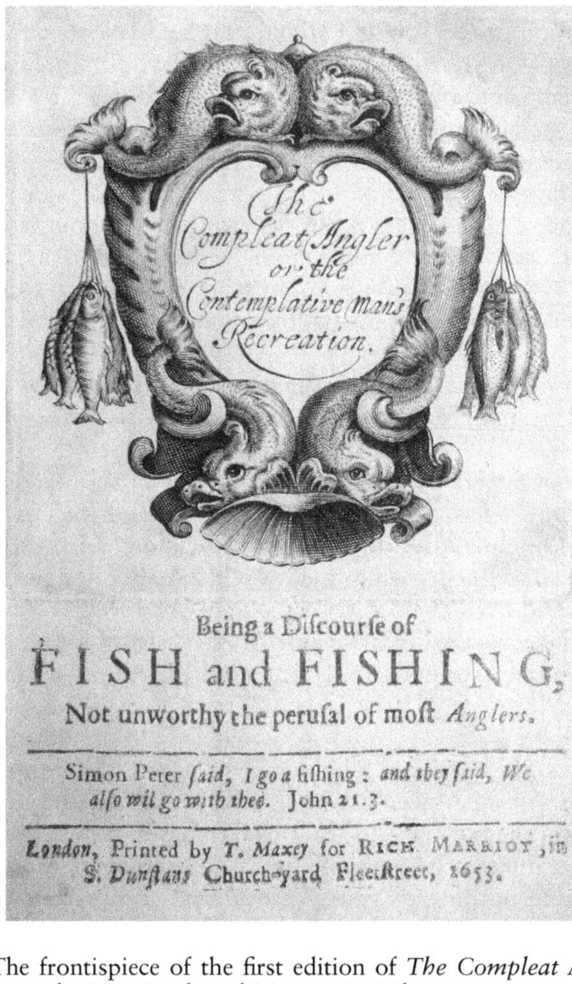
The Compleat Angler or the Contemplative man's Recreation.

Being a Difcourfe of
FISH and FISHING,
Not unworthy the perufal of moft *Anglers*.

Simon Peter *faid, I go a fifhing: and they faid, We alfo wil go with thee.* John 21.3.

*London*, Printed by *T. Maxey* for RICH. MARRIOT, in S. *Dunftans* Church-yard Fleetftreet, 1653.

*Figure 6.2* The frontispiece of the first edition of *The Compleat Angler*, 1653. Beinecke Rare Book and Manuscript Library.

can never be fully learnt; at least, not so fully, but that there will still be more new experiments left for the tryal of other men that succeed us."[4] The repeated call for empirical study corresponds with Walton's frequent invocation of Francis Bacon, whose insistence on "the true proceeding towards active and operative learning" meant that the researcher's methods of observation should be regarded as subject to change in their own right. By Bacon's reckoning, "It is natural to the arts to be in perpetual agitation and growth."[5] Accordingly, Walton's text presents angling as a growing, changing art that is built on the model of learning by doing. After providing instructions on how to gather caddis larvae with a notched stick, Piscator reminds his pupil that listening to others' methods is by no means the best way to learn: "These, my honest Scholer, are some observations told to you as they now come suddenly into my memory, of which you may make some use: but for the practical part, it is that that makes an Angler; it is diligence, and observation, and practice that must do it" (233–34).

At the same time, Walton echoes Bacon in his yoking of empirical study with a vision of collaborative discovery. According to Barbara J. Shapiro, Bacon's approach to the description of non-human phenomena was founded on his experience of legal proceedings. She explains that Bacon was, as a judge, "ideally situated to appropriate legal-historical methods to the cause of natural philosophy."[6] And his approach caught on:

> Bacon's combination of the historico-legal "fact" of human action with the natural fact established by observation and experiment made it possible for his successors to apply a familiar legal technique of verifying events in the human world to natural phenomena. In natural history as in law, "matter of fact" could best be established by the testimony of a sufficient number of firsthand witnesses of appropriate credibility. The term "fact" or "matter of fact" implied, for most of the seventeenth century, not something already worthy of belief or true, but rather a matter capable of proof, preferably by multiple eyewitness testimony.[7]

Walton's text often demonstrates confidence in the idea that an assertion can be proven in the field if the source of the information is not sufficiently trustworthy. At several junctures, Piscator hesitates to make an absolute statement of fact, citing his lack of certainty. This manoeuver often follows the seemingly confidential sharing of another angler's "secret": "This I heard not long since from a friend, but have not tryed it; yet I grant it probable" (139). Having delivered advice from an unknown source about the viability of particular dead baits for catching pike, Piscator hedges on the matter of the information's reliability, then offers up something about which he is totally confident—a recipe: "This [bait] has not been tryed by me, but told me by a friend of note, that pretended to do me a courtesie: but if this direction to catch a Pike thus do you no good, I am certaine this direction how to roste him when he is caught, is choicely good, for I have tryed it" (157). Although both sets of information are secondhand, the recipe being attributed in subsequent editions to an "M. B.," the speaker only fully trusts those instructions that he has tested in his own experiments, and he ranks the directions accordingly.[8]

The privileging of firsthand experience is enforced by the dramatic frame of the text, which foregrounds the Brotherhood of the Angle as an enclave within which investigators can arrive at knowledge *together*. Rather than delivering instructions to imagined readers in his own voice, Walton develops the proxies for teacher and learner that had been introduced by William Samuel in his *Arte of Angling* (1577). In Samuel's treatise, as in the *Angler*, information comes to the reader through the vector of dialogue between a seasoned sportfisher and an initiate.[9] The result, in Walton's text, is that readers can observe the intertwined development of a new friendship and a new set of piscatory skills. As others have noted, there are important similarities and differences between the two treatises. Jonquil Bevan, in her exhaustive edition of the *Angler*, lists several events that Walton borrows from Samuel's dramatized instructions: "The initial demonstration of skill which turns the sceptical Viator into a willing pupil, the loss of a hook (exploited by Samuel as the occasion for some information about hooks, but serving no constructional purpose in the Angler), and the change to a domestic scene in the evening, where the day's catch is produced cooked for supper."[10] With a focus on the emerging friendship among the characters in the *Angler*, Marjorie Swann notes that Walton "departs strikingly from his source" by representing angling "as an inherently gregarious activity." Samuel's fishing instructor is "an acerbic loner who … misanthropically forbids stream-side conversation," but Walton's is overwhelmingly sociable.[11]

## Fishing Buddies

The dialogue doesn't merely present a catechism of questions and answers about how to catch fish; it dramatizes the deliberate actions and subsequent successes of Piscator and his student. The master declares his intent to "catch a Chub … and dress it for our dinner" since "there is not a likely place for a Trout hereabout" (47). He places his student nearby so that the whole proceeding will be observed—a bit of staging that also puts the reader in a specific position relative to fish and angler: "Look you here, Sir, do you see? (but you must stand very close) there lye upon the top of the water twenty Chubs" (48). Viator's expression of doubt is quickly answered by Piscator's landing of the biggest and best fish, and the teacher again uses deictic language to point out his particular success: "Look you, Sir, there he is, that very Chub that I shewed you, with the white spot on his tail" (48). Soon after, the student is given verbal instructions to reinforce what he has witnessed, and is sent off to catch one of his own. He returns triumphant, echoing his teacher's self-praise: "Look you, Master, what I have done; that which joyes my heart; caught just such another Chub as yours was" (54). Viator's announcement invites comparisons between his quarry and his own creaturely heart, both of which have undergone a sudden change.

Rendering the scene in this lively way, with the two characters seeming to show us their catch in real time, Walton achieves the effect of landing the fish directly on the bank of his text, thus blurring the sharp distinction between readerly absorption of information and practical experience in the field. The chances of converting a solitary reader, one who is at home with book in hand rather than out by the river with rod and line, greatly increase when that student seems to receive the very same lesson that "joys the heart" of the blossoming angler in the text. The affective dimension of Viator's learning, inextricable from the intellectual and mechanical know-how that bring the chub to land, becomes even more important as the story progresses. While Walton is careful to show that a heartfelt zeal for the sport is not tantamount to expertise—it's practice that makes perfect—the *Angler* positions affection as the necessary foundation for learning. The word *love*, already liberally deployed throughout the first edition of 1653, gets added to Piscator's congratulatory statement in the book's third edition: "I now see, that with advice and practice you will make an Angler in a short time. Have but a love to it and Ile warrant you."[12] As a beginner, Viator needs a preexisting fondness for the sport. Over time, he will grow "more and more in love with angling" (212). But affection does not move only in this direction, from student to sport. It is also an inherent part of the human relationships that develop by the riverside. Piscator proudly declares, "We anglers all love one another" (5). According to the progression of events in the *Angler*, a student's predilection can only flourish into skill in the context of a powerful new friendship. The love that springs up between Piscator and Viator is instrumental in cultivating and exploiting the novice angler's delight in the sport.

The weather is on the side of the growing affinity between the two men. As if to echo Walton's hope that the reader of the *Angler* will be granted "a rainy evening to read this book in" (A7v), rain pours intermittently throughout the characters' adventures, often causing them to retreat from the riverbank and converse for stretches of time. Like the fishing demonstrations themselves, these waiting sessions bring the two men into close physical proximity and greater familiarity with one another. The episodes are situated not only in a specific meteorological context, but are further defined by the botanical enclosures that keep the anglers in relative dryness and comfort—first a hedge, perhaps made up of honeysuckle bushes (62), and later the canopy of a sycamore tree (114, 121).[13] Piscator introduces these retreats as pauses from the active pursuit of fish, but fills the time with his continued discourse. In fact, when the rain interrupts an effort to catch trout, the fellows leave their rods behind and let the angling proceed without them, so that the sport itself (not just its description) can carry on. Having been tutored in methods for catching various species, Viator urges his teacher to extend his verbal instructions: "Nay, good Master, one fish more, for you see it rains still, and you know our Angles are like mony put to usury; they may thrive though we sit still and do nothing, but talk & enjoy one another" (183).

Mutual enjoyment is evidently fostered in the alternative recreational space provided by trees and shrubs. Piscator compares the shelter of the sycamore to the shade of the beech under which a shepherd and a goatherd converse in Virgil's first Eclogue: "we have ... sate as quietly and as free from cares under this Sycamore, as Virgils Tityrus and his Melibaeus did under their broad Beech tree" (121). But the reference to the pastoral tradition is not without its complications. Marjorie Swann points out that Virgil's herdsmen are not as carefree as Piscator suggests—the displaced Meliboeus may have been read as a reflection of the many English royalists who had recently suffered exile or the seizing of their estates.[14] A more apt Virgilian analogue might have been the retreat of Dido and Aeneas into a cave in Book 4 of the *Aeneid*; that couple's hunting expedition, like the piscatory adventure of Walton's anglers, gets interrupted by a rainstorm, and the site of their refuge is also the place where their shared recreation takes a new and decisive form. The episode was especially familiar at the time of Walton's writing, as the epic had recently appeared in several new translations.[15] If he was thinking dramatically, though, Walton may have had Marlowe's tragedy in mind. In *Dido, Queen of Carthage* (first published in 1594), the lovestruck queen expresses gratitude for the "kind clowdes that sent forth such a curteous storme" (3.4.55), while others are struck by the strangeness of its onset: "Did ever men see such a sudden storme? / Or day so cleere so suddenly orecast?" (4.1.1-2).[16] Dido seizes the opportunity to bring Aeneas into a nearby cave; Piscator draws his new friend into the shelter of a neighboring tree.

For the 1676 edition of the *Angler*, Walton made the sportfishers' retreat under the sycamore substantially more elaborate, and connected it more explicitly to the botanical roles played earlier in the narrative. Piscator further defines the environment under the tree as a "Primrose-bank," remembering that "on this very grass" he once composed some verse.[17] He goes on to mark the landscape with further memories keyed to specific botanical phenomena, recounting an episode involving a fellow angler and a band of gypsies: "I left this place, and saw a Brother of the Angle sit under that hony-suckle-hedg ... I sate down by him, and presently we met with an accidental piece of merriment."[18] In this version, the specific identification of nearby plant life allows Piscator's anecdote to echo with the anglers' earlier retreat. Soon after the successful chub-fishing of the fourth chapter, when the first rain has begun to fall, Piscator gestures toward the comfort of "yonder high honysuckle hedg": "there we'll sit and sing whilst this showr falls so gently upon the teeming earth, and gives a sweeter smel to the lovely flowers that adorn the verdant Meadows."[19] Coming back to the honeysuckle bower in both body and memory, Piscator confirms its function as a *locus amoenus* where affections are likely to flourish between men who've only recently met.

## No Girls Allowed

In all five early editions of Walton's treatise, the latent eroticism of the sweet-smelling landscape around the anglers is brought to the fore by Piscator's memory of spotting "a handsome Milk-maid" in a nearby field—one whose young age and "smooth song" brought him a "second pleasure" to match the beauty of the pastoral scene (63). This milkmaid and her mother coincidentally appear just after being mentioned, so they are held up for Viator's admiration, too. But the mutual attraction of the two men, so clearly sanctioned by their environment, is not eclipsed by the sudden presence of these female characters. While the dairywomen entertain the anglers with food and song, they are obviously not eligible for membership in the Brotherhood of the Angle. The training Viator receives, and the conversion he subsequently undergoes, are gendered experiences that are not available to women.

The gendering of angling is an innovation on Walton's part. While many earlier texts seem to assume a male readership, there are many representations of female anglers in early modern literature. As previous chapters have shown, Plutarch delivers an angling Cleopatra who maintains her sport in Shakespeare's drama, and Lady Mary Wroth stocks her romance with female fishers. Among the commercial fishers of Phineas Fletcher's *Piscatorie Eclogs* is just one character who seems to angle for recreation: the nymph Nicaea.[20] Beyond these literal depictions of women who fish, there are countless female voices who use the matter of angling metaphorically with an authority that is taken for granted. Bellafront, the titular "honest whore" of Dekker and Middleton's play, describes a customer as a gullible catch that only needs to be hauled into her company: "So: giue the fresh Salmon lyne now: let him come a shoare, hee shall serue for my breakefast, tho he goe against my stomack."[21] The unlucky speaker of Isabella Whitney's "admonition," as we have seen, is familiar with the experiences of both angler and fish. There is a precedent for Walton's assumption that most anglers will be men, but not for his precise establishment of a *brotherhood* that so explicitly excludes women.

In the first edition of the *Angler*, gender difference is enforced through the contrast between the leisurely pursuits of the wandering anglers, who wend their way from place to place, and the labor of the hostess and dairywomen, who can be found at particular locations because of their daily domestic work. Although Piscator's reference to Virgil's *Eclogues* may be rather vexed, he does much to make the scene of angling correspond with the pastures of standard bucolic fantasy. As they ramble to and fro enjoying their free time and fresh air, Walton's anglers resemble the barely working shepherds of the received pastoral tradition. Wendy Wall argues that readings of early modern pastoralism ought to attend to "the difference between leisurely gentlemen shepherds and female dairy workers or housewives, differences that might skew the cultural meaning of bucolic imagery." The work of milk-maids, while it may be admired and eroticized in the fictions of the period, was not an apt representation of lighthearted busywork; instead, "dairying emphasized an arduous manual labor at odds with courtly taste." It appears in renderings of the pastoral precisely because of its *contrast* with the fantasy of casual shepherding.[22] Similarly, when Walton's anglers meet up with the dairywomen or arrive at the hostess's lodging, they enjoy the products and services that these women can provide, including the spectacle of their labor. The men's leisure is made all the sweeter by being juxtaposed against the necessary tasks of the women they meet.

The class distinction that so troubles earlier angling writers, who seek to protect the sport from less gentle practitioners, is thus overlaid onto gender in Walton's treatise. While the dialogue includes no verbalized jealousy of the kind that guards the sport from interlopers in other works, Piscator reveals clear expectations as to who catches and who cooks, and the narrative rewards these expectations. Just as there is nothing to definitively exclude women from the company of anglers in Walton's sources, there is nothing that keeps men from cooking. In Thomas Barker's *Art of Angling*, a text to which Piscator sometimes refers,[23] braggery about riverside success is matched with claims of culinary prowess. Barker devotes several pages to cooking demonstrations—not mere recipes like those mentioned or quoted in the *Angler*, but step-by-step instructions based on the author's "seven times seven years" of experience serving in distinguished English kitchens.[24] Walton's Piscator sounds like Barker when he gloats to Viator about his method for preparing chub—"Though a Chub be by you and many others reckoned the worst of all fish, yet you shall see I'll make it good fish by dressing it" (48)—but his culinary work ends at the delivery of instructions to their hostess at the alehouse: "Come Hostis, how do you? wil you first give us a cup of your best Ale, and then dress this Chub, as you drest my last, when I and my friend were here about eight or ten daies ago? but you must do me one courtesie, it must be done instantly" (50). In Barker, cookery is a natural extension of the angler's effort; in Walton, it's a chore that can easily be delegated to one who is presumably more adept at it.

The 1655 edition of the *Angler*, with its many amplifications, makes specific mention of women anglers. Ironically, these few references serve to reinforce a distinction between the all-male realm of serious sportfishing and a separate, seasonal habit that is observed in women and youth. Cooking is again introduced as a matter of interest, and again categorized as feminine activity. In a new section on the minnow, Piscator explains:

> He is a sharp biter at a small worm, and in hot weather makes excellent sport for young Anglers, or boyes, or women that love that Recreation, and in the spring they make of them excellent Minnow Tansies; for being washed well in salt, and their heads and tails cut off, and their guts taken out, they prove excellent for that use, being fryed with yolks of eggs, the flowers of Cowslips, and of Primroses and Tansie.[25]

The presumptive reader—male, adult—who is enticed by this casual recipe for a flowery minnow omelet may set down his book and head to the river for an easy catch, but if he does so, he will be engaging in behavior more appropriate to a child or woman. Piscator does not position his student in the category of fair-weather anglers who ought to pursue the most gullible fish on the sunniest days. He expects Viator (and, by extension, the reader) to take up greater challenges, so the details about the life histories of small, easily caught species are unapologetically omitted. Speaking of the "sticklebag," Piscator admits, "I know not where he dwels in winter, nor what he is good for in Summer, but only to make sport for boyes and women Anglers, and to feed other fish that be fish of prey, as Trouts in particular."[26] The piscatory endeavors of women and boys are on a level with the predatory appetite of the trout: not practiced skills, but self-feeding activities that become more common as the weather warms in the summertime.

"Boyes and women Anglers," set apart by their age and gender, practice a wholly different sport from the art into which Piscator inducts his student. Within their marginal demographic, only the young males can develop into full-fledged members of the Brotherhood of the Angle. Such a transformation is narrated in the *Angler* itself, through the tutelage and conversion of Viator. The impossibility of a woman becoming such a member is so obvious as to be treated as a biological fact, while a man's transition from non-angler to angler is similarly a consequence of environmental pressures and opportunities. Cast in terms of cognitive ecology, the male prospective angler exhibits a particular kind of plasticity, a propensity to learn and alter his behavior, that is not observed in women. Furthermore, his development is not only a conversion but a metamorphosis—a change from one way of being to another, radically different form. We don't see the shape of Viator drastically altered, but we watch him undergo a process of indirect development that feels to him like a change in kind.

As the newly acquainted travelers approach the thatched house where they are to stay the night, Piscator suggests that they raise a glass to the otter hunters who are planning to meet Viator the next morning. Viator reciprocates: "That we wil, Sir, and to all the lovers of Angling too, of which number, I am now one my self, for by the help of your good discourse and company, I have put on new thoughts both of the Art of Angling, and of all that profess it" (39). At this juncture, the initiate can call himself a lover of the sport, but not yet an outright angler. When he finally bids his teacher farewell at the end of the treatise, Viator will feel himself truly transformed: "Your company and discourse have been so pleasant that, I may truly say, I have only lived, since I enjoyed you and them, and turned Angler" (246). The verb *to turn* in Viator's formulation corresponds with both conversion—literally, a turning toward[27]—and biological metamorphosis.

## Turning Angler

At many stages in the instructional program of the *Angler*, Piscator speaks of creatures that *turn*, changing from one form to another or quickening into life from seemingly inanimate matter. In this way, Walton adopts his sources' enthusiasm for larval forms, dwelling on the mysteries of how and when an individual life begins or undergoes some drastic alteration. Reflecting upon Edward Topsell's descriptions of insect origins,[28] Piscator mentions that certain caterpillars "in their time turn to be Butter-flies; and again, that their eggs turn the following year to be Caterpillers" (99–100). He claims that he has beheld some such transformations: "It is to me observable that at a fix'd age this Caterpiller gives over to eat, and towards winter comes to be cover'd over with a strange shell or crust, and so lives a kind of dead life, without eating all the winter, and (as others of several kinds turn to be several kinds of flies and vermin, the Spring following) so this Caterpiller then turns to be a painted Butterflye" (101–2). By way of "a little confirmation," Piscator then offers several verses from Guillaume Du Bartas's rendering of the creation story, *La Sepmaine, ou Création du Monde* (1578). Walton's excerpt is copied imperfectly from the 1608 edition of Josuah Sylvester's English translation: "God … made many creatures breed / Of liveless bodies, without Venus deed" (102).[29]

The examples borrowed from Du Bartas include transformations of dead plant matter into living animal forms. There are reports of "Goslings hatcht of trees, / Whose fruitful leaves falling into the water, / Are turn'd ('tis known) to living fowls soon after," and "Barnacles" (arctic geese) who grow from "rotten planks of broken ships": "Oh transformation strange! / 'Twas first a green tree, then a broken hull, / Lately a Mushroom, now a flying Gull" (103).[30] In these accounts, progeny spring up where they are least expected, seemingly fostered by radically different species and undeterred by the boundary that separates life from death. Walton's Piscator explicitly associates these stories of asexual reproduction with the life cycles of familiar insects, comparing the "dead life" of pupating caterpillars to the mysteriously "fruitful" properties of decaying leaves and timber. The *turning* of larval forms into adult creatures—the emergence of the butterfly into what Piscator might call "living life"—is tantamount to spontaneous generation.

When he ponders the mysteries of eel reproduction later in the dialogue, Piscator cautiously returns to Du Bartas's examples of avian offspring. Considering the hypothesis "that Eeles are bred of a particular dew falling in the Months of May or June on the banks of some particular Ponds or Rivers (apted by nature for that end) which in a few dayes is by the Suns heat turned into Eeles" (188), the master angler withholds an outright confirmation, noting only that he has seen young eels covering parts of a river's surface and heard of other waterways where such profusion appears. But he goes on to say that spontaneous generation "seems to be made more probable by the Barnacles and young Goslings bred by the Suns heat and the rotten planks of an old Ship, and hatched of trees," citing Du Bartas and other authorities who present these cases "for truths" (189). Piscator is hesitant to advance a theory for which he has little firsthand evidence, but he is quick to compare the larval stages of various species and to suggest that the time before life—or between one mode of living and the next—is mysterious in all kinds of creatures.

As in earlier treatises, the subject of larvae in the *Angler* gets its most thorough treatment in instructions about how to gather and maintain live bait. Walton has Piscator advise his student to procure a cat or kite carcass and let it become fly-blown, so that viable bait will readily appear. The corpse can be buried and its resident maggots harvested whenever they are needed, at least until such time as they "turn to be flies" in the spring (224). That metamorphosis is the second observable bodily change Piscator's "gentles" undergo; the first occurs when they "begin to be alive and to stir" (224). At the end of the narrative, Viator will echo this language as he attests to having quickened under the influence of the master sportfisher:

> I thank you for your many instructions, which I will not forget; your company and discourse have been so pleasant, that I may truly say, I have only lived, since I enjoyed you and them, and turned Angler. I am sorry to part with you here, here in this place where I first met you, but it must be so: I shall long for the ninth of May, for then we are to meet at Charls Brandons. This intermitted time wil seem to

> me (as it does to men in sorrow) to pass slowly, but I wil hasten it as fast as I can by my wishes, and in the mean time the blessing of Saint Peters Master be with mine.
>
> (245–46)

Looking forward to their eventual meet-up at the tackle shop but already preparing for a period of lonesome dormancy, Viator sounds not only like a man "in sorrow" but a man in love. The change he feels in himself is made more acute by the anglers' having circled back to the site of their first encounter.

In the midst of such sentimentality, we can nevertheless recognize that the tutelage of a novice angler and the farming of maggots are parallel pursuits—Piscator's carefully cultivated bait, like the piscatory skill of the neophyte, can come into being only in a particular environment. The angler must enact assiduous care for the development of the nascent creature, and must choose the conditions that are most advantageous for the changes ahead. In the case of Viator and of many insect species, those conditions include specific botanical habitats. Without the honeysuckle hedge or the sycamore tree, Piscator would lack a nursery for his not-yet-angler. Just as "the hawthorn fly is to be had on any Hawthorn bush, after the leaves be come forth" (118), so is Viator to be found at his lessons under the sycamore tree while the rain pours.

Viator's change is brought about primarily through learning; he cannot become an angler without both listening to the lengthy didactic speeches of his new friend and putting those lessons into practice. The practical dimension of his training is not immediately available to the curious reader, but the text returns again and again to the notion of learning by doing. In his own voice and through the persona of Piscator, Walton promotes experimentation in the field as the only sure way to gain expertise. Like larval invertebrates who must spend plenty of time and energy in order to emerge from their "dead life" into new bodily forms, the student of angling is not fully *turned* until he rises up from the realm of discourse and actively applies his knowledge at the riverside. The time spent under a sycamore tree with a wise tutor (or at home with an open copy of *The Compleat Angler*) is comparable to the "larval diapause" taken by many insects: "a period of developmental and physiological quiescence as an adaptation for surviving seasonally recurring adverse conditions."[31] By comparison with what follows it, the angling student's quiet absorption of information is a phase of relative inactivity. It is encouraged, if not necessitated, by meteorological conditions, and it ends with a morphological change: the angler's body is extended when he takes up a rod and steps into his new identity. He has also clearly undergone an internal change that will determine how he lives and thinks in the future.

Because learning involves a considerable investment of time and energy, cognitive ecologists have seen fit to ask what purpose it serves.

A period of learning in an animal's development can come at a steep cost—juveniles are likely to suffer drastically higher mortality rates than their adult conspecifics because they lack sufficiently sophisticated feeding techniques and haven't yet acquired the necessary patterns of antipredatory behavior.[32] However, as Dukas observes:

> The ecological cost of inexperience … has to be viewed in perspective: learning allows animals to exploit environmental features, behavioral repertoire, and niches that cannot be used otherwise. So the initial period of inexperience can be seen as a necessary component of a life history that relies on and ultimately benefits from learning.[33]

Furthermore, from studies conducted on fruit flies, we now know that learning can take place before metamorphosis. In the lab, fruit fly larvae have been shown to choose an odor that has no negative association over one that is associated with an electric shock. This capacity is surprising, given that all fruit flies need to do when they hatch is to begin to feed, right on their home turf of yeast-rich rotting fruit—"even though the larval task seems easy and they possess limited sensory and information-processing abilities, larvae are able to learn."[34] In the nursery of the wet, green riparian zone, Walton's emergent angler is also in a position to absorb and use new information, developing an inward competence that will eventually facilitate his full transformation.

## Studied Conversion

Walton approaches the question of learning and its developmental functions in his *Life of Donne*, too, where the story of the dean's youth is particularly problematic because of his Catholic upbringing. Walton presents but quietly dismisses the idea that Donne was "another Picus Mirandula; of whom Story sayes, That he was rather born than made wise by study."[35] Instead of promoting the notion of Donne as a prodigy for whom the usual educational avenues would have been redundant, Walton is keen to show that the boy was deliberately schooled in various environments, beginning at home in his early childhood. After receiving private tutoring, the ten-year-old Donne goes off to Oxford and then, at age fourteen, to Cambridge, "that he might receive nourishment from both Soiles." He is not a prodigy but a "most laborious Student" at Cambridge, until his admission at seventeen to Lincoln's Inn. Money left by his now deceased father allows Donne's mother to employ a new complement of private Catholic tutors to instruct the young man in mathematics, "Liberall Sciences," and theology.[36] His aptitude for higher learning is met with a financial and spiritual investment on the part of his parent, so that the young scholar can indeed be "made wise by study."

Like his learning more generally, Donne's conversion is not a matter of acquiescence to natural influences but an intellectual undertaking that involves individual study, the gathering of resources, and discourse with other parties (including God). The transition to Protestantism is hard work, inspired by a gap that the young man perceives in his own worldview; because he is "unresolved in his Religion" he must "survey the Body of Divinity" and make a decision about the proper orientation of his faith.[37] The bright light of truth is, in Walton's account, on the side of the reformed church, but the rightness of Protestantism is just one factor in Donne's eventual change; he arrives at truth only through study. As a larval Protestant minister who begins life as something much different, the young Donne is not wholly unlike the potential angler who emerges into a new form by the end of Walton's fishing treatise. Both arrive at a fundamental truth that has been advanced by other voices. Both require a literal change of scenery and exposure to expertise. And both, because these are conversion narratives, become zealous advocates for the new way of life into which they have been reborn.[38]

Molly Murray explains that early Christian conceptions of conversion defined the experience in two simultaneous ways, "as a change of church and as a change of soul;" to convert to Christianity was not only to openly perform certain rituals, but also to change inwardly. The name given to this less visible alteration was *metanoia*, "a word that literally indicates a change in spirit or mind."[39] *Metanoia* can easily be applied to the silent changes that occur within the developing angler. Such changes are observable in Viator's statements about his shifting orientation: the joy that fills his heart when he catches his first chub, the sorrow he feels at parting ways with his beloved angling-master, and the sensation he has of emerging into life for the first time. Moreover, religious conversion itself is a subject that arises within the narrative frame of the *Angler*: Piscator borrows from Donne the observation that Solomon "before his conversion was remarkably amorous, and after by Gods appointment, writ that Love-Song betwixt God and his Church" (29).[40] And as several readers have noted, Walton appears to pun on his own religious affiliation and the name of the sport—the "compleat angler" is also a complete Anglican.[41] But the text does not linger in solemn reflection, nor does it make angling an opportunity for solitary devotional practice, as several of its sources do. There is only one community into which Walton earnestly endeavors to draw us, and that is the Brotherhood of the Angle.

The notion of achieving conversion through text reaches back, as Murray shows, to Augustine's epiphanic reading, "a private hermeneutic quickening that would inspire generations of Christian theologians." In early modern England, that concept was renewed through the story of the Protestant martyr Anne Askew; setting down their accounts of her life and death, John Foxe and John Bale adopted "the familiar idea of conversion as a process of repentance inspired by reading."[42] As a text that repeatedly refers to its own inadequacy, urging its readers to venture outdoors and get close to some real experts, Walton's *Angler* is painfully aware of its limitations. But it's instructional in another way; it presents an author who knows he is second-best, an angler but an "unlearned" one, and therefore an authority on how a person might come to number among the brothers of the angle. In order to frame a treatise that is "not unworthy the perusal of most Anglers," Walton has to transform himself into the more seasoned Piscator, a character who ultimately exhibits much of his creator's hesitation and humility anyway. The self-consciousness of the author, from the prefatory matter onward, helps readers recognize that their guide is undergoing a change in his own right. Walton is becoming an expert angler as he reads and writes.

Fittingly, the *conversions* that the anglers mention, setting aside the example of Solomon, are transmutations of textual matter from one form to another. Piscator explains that he has "taken a little pleasant pains to make … a conversion" of an epigram about scoffers from the dialogues of Lucian (7). The second converted text is Viator's angling song; he announces near the end of the narrative that he has "made a conversion of a piece of an old Ketch, and added more to it, fitting them to be sung by us Anglers" (215). Although they have obviously sprung up from the labor of Walton and his sources, these adaptations are now doubly or triply converted by being attributed to the fictional characters who display them. As he transforms texts from various quarters, Walton demonstrates his intention to change both himself and his reader. Rather than aiming at the impossible *completion* enshrined in his title, he draws the hopeful angler into a world of perpetual turning, where even the most larval among us may be proven apt learners.

## Notes

1 Izaak Walton, *The Compleat Angler or, the Contemplative Man's Recreation* (1653, Wing [2nd ed.] W661), A3v, A4v, A3r. Unless otherwise indicated, text citations refer to signature or page numbers in this, the first edition.

2 Jonquil Bevan, introduction to *The Compleat Angler 1653–1676*, by Izaak Walton (Oxford: Clarendon Press, 1983), 35.

3 Swann contends, "In *The Compleat Angler* we find the confluence of several facets of early modern collecting: Baconian natural history and the rise of the 'virtuosos'; a premium placed on 'wonder' and 'wonderful' physical objects; the importance of textualization in the process of collecting; and the use of collections to create new social groups" ("*The Compleat Angler* and the Early Modern Culture of Collecting," *English Literary Renaissance* 37, no. 1 [2007], 101).

4 Izaak Walton, *The Compleat Angler*, 2nd ed. (1655, Wing [2nd ed.] W662), A6v.

5 Francis Bacon, "Filum Labyrinthi," in *The Works of Francis Bacon*, ed. James Spedding, Robert Leslie Ellis, and Douglas Denon (Boston: Houghton Mifflin, n.d., ca. 1900), 6:427, 425. Walton makes many references to Bacon's unfinished natural history, the first volume of which was published in 1622 (*Historia naturalis et experimentalis ad condendam philosophiam; sive phaenomena universi*). Walton most often draws information from the second part, *Historia vitae et mortis* (History of Life and Death), which came out in 1623; see Markku Peltonen, "Bacon, Francis, Viscount St Alban (1561–1626), Lord Chancellor, Politician, and Philosopher," *DNB* (October 2007), https://doi.org/10.1093/ref:odnb/990.
6 Barbara J. Shapiro, *A Culture of Fact: England, 1550–1720* (Ithaca, NY: Cornell University Press, 2000), 108.
7 Shapiro, "A Culture of Fact," 109–10.
8 "M.B." has not been identified, but Jonquil Bevan notes that the recipe adopted by Walton is fairly standard (*The Compleat Angler 1653–1676*, 405n290:30).
9 [William Samuel], *The Arte of Angling* (1577), ed. Gerald Eades Bentley (Princeton, NJ: Princeton University Press, 1956). The names Piscator and Viator come from Samuel's treatise, too.
10 Bevan, introduction to *The Compleat Angler 1653–1676*, 17.
11 Marjorie Swann, introduction to *The Compleat Angler*, by Izaak Walton and Charles Cotton (Oxford: Oxford University Press, 2014), xiii.
12 Walton, *The Compleat Angler*, 3rd ed. (1661, Wing W663), 61.
13 The hedge is given no botanical specificity in the early editions of the *Angler*, but Piscator calls it "yonder high honysuckle hedg" in the 1676 edition (Wing [2nd ed.] W666, 78). The sycamore provides both dry shelter from the rain and cool shade from the sun. While the weather is good, the anglers take their breakfast under it (104, 107).
14 Swann, introduction to *The Compleat Angler*, xvi.
15 Sheldon Brammall examines the various motivations of Virgil's early modern English translators in his chronological study *The English Aeneid: Translations of Virgil 1555–1646* (Edinburgh: Edinburgh University Press, 2015).
16 Christopher Marlowe, *Dido, Queen of Carthage*, in *The Complete Works of Christopher Marlowe*, 2nd ed., vol. 1, ed. Fredson Bowers (Cambridge: Cambridge University Press, 1981).
17 Walton, *The Compleat Angler*, 5th ed. (1676, Wing [2nd ed.] W666), 123.
18 Walton, *The Compleat Angler*, 5th ed., 124.
19 Walton, *The Compleat Angler*, 5th ed., 78.
20 Phineas Fletcher, *The Purple Island or the Isle of Man together with Piscatorie Eclogs and other Poeticall Miscellanies* (1633, STC [2nd ed.] 11802), 30.
21 Thomas Dekker and Thomas Middleton, *The Honest Whore, With the Humours of the Patient Man, and the Longing Wife* (1604, STC [2nd ed.] 6501), C4v.
22 Wendy Wall, *Staging Domesticity: Household Work and English Identity in Early Modern Drama* (Cambridge: Cambridge University Press, 2002), 134.
23 Piscator's explicit references to Barker can be found on pages 108 and 116 of the first edition of *The Compleat Angler*.
24 Thomas Barker, *The Art of Angling* (1653, Wing B783), 17.
25 Walton, *The Compleat Angler*, 2nd ed., 320.
26 Walton, *The Compleat Angler*, 2nd ed., 324. Piscator refers to either the three-spined stickleback (*Gasterosteus aculeatus*) or the nine-spined stickleback

(*Pungitius pungitius*). Both are tiny, rarely growing beyond seven centimeters in length (Malcolm Greenhalgh, *The Pocket Guide to Freshwater Fish of Britain and Europe* [London: Mitchell Beazley, 2001], 156–59).

27 *OED*, s.v. "conversion, n.," sense 2, accessed April 25, 2018, www.oed.com/view/Entry/40773?redirectedFrom=conversion.

28 Edward Topsell, *The Historie of Serpents* (1608, STC [2nd ed.] 24124), 106–8.

29 Josuah Sylvester, *Bartas his Deuine Weekes and Workes* (1608, STC [2nd ed.] 21650). Sylvester's translation brings together the creation story (the first week) and a subsequent installment of biblical narrative (the second week). Swann notes that the verses in question are from the fifth day of the first week (lines 29–48) (*The Compleat Angler*, 242n34).

30 *OED*, s.v. "barnacle, n.2," accessed April 26, 2018, www.oed.com/view/Entry/15624?rskey=0yD1yp&result=2&isAdvanced=false.

31 H. Frederik Nijhout, "Hormonal Control in Larval Development and Evolution—Insects," in *The Origin and Evolution of Larval Forms*, ed. Brian K. Hall and Marvalee H. Wake (San Diego, UK: Academic Press, 1999), 240.

32 Reuven Dukas, "Learning: Mechanisms, Ecology, and Evolution," in *Cognitive Ecology II*, ed. Reuven Dukas and John M. Ratcliffe (Chicago: University of Chicago Press, 2009), 15.

33 Dukas, "Learning," 16.

34 Dukas, "Learning," 17.

35 Izaak Walton, *The Life of John Donne, Dr. in Divinity, and Late Dean of Saint Pauls Church London* (1658, Wing W668), 6. This is the second edition of Walton's biography of Donne; the first was printed as the preface to Donne's sermons in 1640 (Jessica Martin, *Walton's* Lives*: Conformist Commemorations and the Rise of Biography* [Oxford: Oxford University Press, 2001], 168).

36 Walton, *The Life of John Donne*, 7–8.

37 Walton, *The Life of John Donne*, 10.

38 In a chapter on "John Donne and the language of de-nomination," Molly Murray bucks the trend of "reading Donne's devotional poetry as clear statements of a broad, bland middle way." Instead, she argues for "taking Donne at his word: that he was born into the Roman Catholicism of his family, and that he converted to the Protestantism of his nation. This conversion seems to have been sincere, and it was certainly final." In *The Poetics of Conversion in Early Modern English Literature* (Cambridge: Cambridge University Press, 2009), 71.

39 Murray, *The Poetics of Conversion*, 7.

40 Swann, notes to *The Compleat Angler*, 243n38.

41 Swann, introduction to *The Compleat Angler*, xi–xii.

42 Murray, *The Poetics of Conversion*, 10, 16.

## References

Bacon, Francis. "Filum Labyrinthi." In *The Works of Francis Bacon*, edited by James Spedding, Robert Leslie Ellis, and Douglas Denon. Boston: Houghton Mifflin, n.d., ca. 1900. 6:413–28.

———. *Historia naturalis et experimentalis ad condendam philosophiam; sive phaenomena universi*. London, 1622. STC (2nd ed.) 1155.

———. *Historia vitae et mortis*. London, 1623. STC (2nd ed.) 1156.

Barker, Thomas. *The Art of Angling*. London, 1653. Wing B783.

Bevan, Jonquil. Introduction to *The Compleat Angler 1653–1676*, by Izaak Walton. Oxford: Clarendon Press, 1983.

Bramall, Sheldon. *The English Aeneid: Translations of Virgil 1555–1646*. Edinburgh: Edinburgh University Press, 2015.

Dekker, Thomas and Thomas Middleton. *The Honest Whore, With the Humours of the Patient Man, and the Longing Wife*. London, 1604. STC (2nd ed.) 6501.

Du Bartas, Guillaume de Salluste. *La Sepmaine, ou Création du monde*. Paris, 1578.

Dukas, Reuven. "Learning: Mechanisms, Ecology, and Evolution." In *Cognitive Ecology II*, edited by Reuven Dukas and John M. Ratcliffe, 7–26. Chicago: University of Chicago Press, 2009.

Fletcher, Phineas. *The Purple Island or the Isle of Man together with Piscatorie Eclogs and other Poeticall Miscellanies*. Cambridge, 1633. STC (2nd ed.) 11802.

Greenhalgh, Malcolm. *The Pocket Guide to Freshwater Fish of Britain and Europe*. London: Mitchell Beazley, 2001.

Marlowe, Christopher. *Dido, Queen of Carthage*. In *The Complete Works of Christopher Marlowe*, edited by Fredson Bowers (2nd ed., vol. 1), 1–252. Cambridge: Cambridge University Press, 1981.

Martin, Jessica. *Walton's Lives: Conformist Commemorations and the Rise of Biography*. Oxford: Oxford University Press, 2001.

Murray, Molly. *The Poetics of Conversion in Early Modern English Literature*. Cambridge: Cambridge University Press, 2009.

Nijhout, H. Frederik. "Hormonal Control in Larval Development and Evolution—Insects." In *The Origin and Evolution of Larval Forms*, edited by Brian K. Hall and Marvalee H. Wake, 217–54. San Diego, CA: Academic Press, 1999.

Peltonen, Markku. "Bacon, Francis, Viscount St Alban (1561–1626), Lord Chancellor, Politician, and Philosopher." In *Oxford Dictionary of National Biography*. Oxford University Press, October 2007. https://doi.org/10.1093/ref:odnb/990.

[Samuel, William]. *The Arte of Angling*. 1577. Edited by Gerald Eades Bentley. Princeton, NJ: Princeton University Press, 1956.

Shapiro, Barbara J. *A Culture of Fact: England, 1550–1720*. Ithaca, NY: Cornell University Press, 2000.

Swann, Marjorie. "*The Compleat Angler* and the Early Modern Culture of Collecting." *English Literary Renaissance* 37, no. 1 (2007): 100–17.

———, ed. Introduction to *The Compleat Angler*, by Izaak Walton and Charles Cotton. Oxford: Oxford University Press, 2014.

Sylvester, Josuah. *Bartas his Deuine Weekes and Workes*. London, 1608. STC (2nd ed.) 21650.

Topsell, Edward. *The Historie of Serpents*. London, 1608. STC (2nd ed.) 24124.

Wall, Wendy. *Staging Domesticity: Household Work and English Identity in Early Modern Drama*. Cambridge: Cambridge University Press, 2002.

Walton, Izaak. *The Compleat Angler or, the Contemplative Man's Recreation*. 1st ed. London, 1653. Wing (2nd ed.) W661.

———. *The Compleat Angler*. 2nd ed. London, 1655. Wing (2nd ed.) W662.

———. *The Compleat Angler*. 3rd ed. London, 1661. Wing W663.

———. *The Compleat Angler*. 5th ed. London, 1676. Wing 666.

———. "The Life and Death of Dr Donne, Late Deane of St Pauls London." Preface to *LXXX sermons preached by that learned and reverend divine, Iohn Donne, Dr in Divinity, late Deane of the cathedrall church of S. Pauls London*, by John Donne. London, 1640. STC (2nd ed. 7038).

———. *The Life of John Donne, Dr. in Divinity, and Late Dean of Saint Pauls Church London*. London, 1658. Wing W668.

# Conclusion

## Killing and Conservation

The irony at the heart of angling is never far from the minds of early modern authors when they use the sport to formulate ideas about human relationships. Nor is it absent from descriptions of the angler's practice in the fishing manuals of the sixteenth and seventeenth centuries, where the manipulation of piscine hunger receives careful, species-by-species attention. The enduring conceptual metaphors that yield expressions like "fishing for compliments" and "falling hook, line, and sinker" are based on the obvious fact that fish do not choose to be caught. Real and imagined anglers in early modern sources are keenly aware of the differing interests that bring humans and fish into contact at the water's edge, and they know that their sport could result in the devastation of species and habitats if it isn't properly managed. From these texts, and from the network of cultural understandings that shaped them, we have inherited a picture of angling as a wild contest in which one subject uses deceit to win the life of another. Attendant upon this standard view, though, is another notion with its own irony: for centuries, sportfishing, like other hunting practices, has been associated with efforts of conservation.

Marjorie Swann interprets the gruesome otter-hunt in *The Compleat Angler* as evidence of Piscator's "concern for wildlife conservation," since the killing of otters is apparently undertaken to protect wild fish populations. Although she notes the paradox at work here, Swann also finds an apt continuity between Walton's orientation toward violence in the *Angler* and the more recent agenda of habitat preservation:

> In twentieth-century America, Walton's advocacy of environmental stewardship inspired the genesis of the Izaak Walton League, a mass-membership conservation organization that was especially influential from the early 1920s until the Second World War. The Izaak Walton League promoted what was at the time a cutting-edge vision of conservation focused broadly on ecosystems, and the group was crucial in forcing Congress to protect endangered watersheds by creating federal nature preserves. Izaak Walton the otter-hater thus became

> a progenitor of the modern American conservation movement, a 'Defender of Woods, Waters and Wild Life'.[1]

If we accept what early modern anglers have to say about their commitment to protecting fish populations, the legacy that connects their sport to a lasting rhetoric of wildlife conservation makes perfect sense. But there are good reasons to take a second look at that established link and ask whether the textual record of angling has left us with other, less vexed ways of thinking about piscine lives and habitats.

After a long career working for the provincial government, British Columbian conservation biologist Dionys de Leeuw took up the ethical question "Can Nature Conservation Justify SportFishing?" In an article under that title, de Leeuw writes in *Environmental Ethics* about the long-standing claims proponents of fishing and hunting have made for their sports on the basis of protecting wild environments.[2] His epigraph comes from the *Treatyse of fysshynge wyth an Angle*, in the modernized English of John McDonald's edition: "Also you should busy yourself to nourish the game in everything that you can, and to destroy all such things as are devourers of it."[3] Reaching back to the advent of piscatory literature in English, de Leeuw reveals with this gesture that the rhetoric of conservation has relied—for a very long time—on an articulated willingness to use violence. There is nothing unusual, in either early modern conceptions of "nourish[ing] the game" or in more recent studies of conservation, about the assertion that one kind of organism must be eradicated in order to maintain a population that is seen as more valuable, vulnerable, or established. De Leeuw calls attention to the injunction to protect the resource by "destroy[ing]" any perceived threats, but he also registers the ambiguity of the word "devourers." Might this category include anglers themselves? How should we distinguish between destruction and devouring?

De Leeuw credits the sport with increasing the store of human knowledge about other species:

> Besides formal studies by management organizations and research institutions, anglers themselves take great interest in many aspects of their sport such as entomology, fish behavior, and general nature study. Angling, therefore, not only generates an enormous amount of information on nature, it also creates an interest in nature.[4]

This *interest* can be read economically, as an investment made by people who wish to know more because they believe that knowledge is a good in itself. But an increased sensitivity to the complexity of the biotic community does not necessarily bring anglers into practices that are less harmful to fish and their habitats. While some management efforts have led to the protection or restoration of ecosystems, others have seriously compromised fish and their neighbor species. There is also the difficulty of defending an elective activity that directly injures or kills individual animals. According to de Leeuw's reckoning, "When harm to fish by angling is included, sport fishing becomes increasingly unjustifiable on the basis of nature conservation."[5] Unsurprisingly, de Leeuw understands *harm* to include both suffering and untimely death. The paradox that he seeks to unravel—that fish must be protected in order to be caught, and caught in order to be protected—is built on an implicit claim of human exceptionalism that has shaped the sport and its poetics. A return to early modern angling literature, in which the violence and death that characterize the aquatic hunt are clearly and unapologetically on display, can expose how the question of conservation has come to eclipse the matter of killing.

Early modern piscatory writers are not at pains to excuse their sport or to extol its virtues as a force for sustainability. What we do see is a habit of situating the fisher's activity in an ecosystem, and an insistence that the contemplation of that broader biotic world is an inevitable and positive consequence of standing by the river, rod in hand, for hours on end. The "interest in nature" generated by the idealized sport is, for many early modern author-anglers, both intellectual and spiritual. A similar case is made in contemporary accounts of angling. In an article for the *New York Times*, the writer and artist James Prosek describes the angler's feeling of being part of a bigger picture: "It is not, ... nor has it ever been, simply about the predation itself, but the whole assemblage of steps before and after, about engaging our senses with a larger interconnected whole (from anticipation and pursuit to the preparation and eventual ingestion of our prey)."[6] Prosek's account is secular, but it shares with early modern angling manuals a kind of faith in the value of considering one's place in the world, in relation to both the immediately observable environment and a much broader imagined ecology.

In fact, faith numbers first among the virtues that an angler must exhibit, according to William Samuel's treatise of 1577, *The Arte of Angling*. For Samuel, this amounts merely to the fisher's "believing that there is fish where he cometh to angle."[7] When John Dennys elaborates upon the qualities in this catalogue, piscatory faith is charged with religious meaning; in *The Secrets of Angling*, the angler has to believe

> That where as waters, brookes, and lakes are found,
> There store of Fish without all doubt abound ...
> For nature that hath made no emptie thing ...
> Hath fild each Brooke, each Riuer, Lake and Spring
> With creatures ...[8]

Faith in abundance looks like faith in human supremacy here, especially considering that Dennys compares the angler's steadfastness to that of Abraham, whose commitment "was so acceptable [that] his increase and of-spring manifolde / Exceeded far the starres innumerable."[9] There are plenty of fish in the river, and plenty of people to catch them, because human beings have a special providential relationship with God. It falls to the hopeful angler only to monitor the climate and make prudent choices about when it's worthwhile to pack up one's gear and venture out for a day of fishing. The speaker of *The Secrets of Angling* reveals a world that is filled with signs for anglers to read.

For example, a sportfisher's practice has to change with the onset of winter, because fish conduct themselves differently then:

> … when the golden Chariot of the Sunne,
> Departing from our Northren countries farre
> Beyond the ballance, now his course hath runne,
> And goes to warme the cold *Antarctique* starre,
> And Summers heat is almost spent and done,
> With new approch of Winters dreadfull warre:
>     Then doe the Fish withdraw into the deepe,
>     And low from sight and cold more close doe keepe.[10]

Dennys follows epic convention in his couching of detailed information about fish behavior in the vast context of celestial movement and seasonal transition. This need to situate angling in a universe of partly predictable earthly conditions is a habit that never falls away from the sport's attendant literature. In piscatory writing of our own time, evidence of the angler's ecological turn of mind lends a certain legitimacy to the claim that the sport is fundamentally compatible with—or essential to—efforts of conservation.

Echoing *The Secrets of Angling*, American fishing journalist Zach Matthews writes about the beginning of the brown trout's reproductive cycle in the Cumberland River:

> When the sun reaches a certain point in its yearly orbit, or when the leaves turn, or the air grows crisp, or the kids break out the football jerseys (no one's really sure what actually triggers the event), the brown trout start to make their way onto the shoals (or "bars"). Usually browns start staging in the Cumberland by mid-September. At some predetermined but unknowable signal, the first females (or "hens") will begin to cut redds (which is what biologists call their nests).[11]

Matthews's purpose in this article is to show readers that it's still possible to enjoy a good day of fishing on the Cumberland. Despite the erosion of a nearby dam, which used to bring gushes of cold water into the river and allow for a "bonanza" of fish every October, anglers can be successful if they adjust their approach and their expectations. He counsels:

> The best thing anglers can do in this time of stress is to practice quick catch-and-release and remain good stewards of the river .... Only by engaging with the river can we hope to teach the anglers of tomorrow what a special thing they have, so that when the time comes, they too will want to preserve it.[12]

Fishers are not fellow creatures but "stewards" of a habitat in which they themselves could never survive; the river is "a special thing they have," like God's promise to Abraham. The idea that the riparian zone and its biota could be protected through the total cessation of fishing just doesn't come up, because aquatic conservation is understood as an effort carried out in the interest of anglers, largely *by* anglers, without whom there would be no reason to mitigate against the negative effects of other human activities on river life.

Despite Prosek's claim that predation is not the sole purpose of angling, and Matthews's promotion of catch-and-release as a viable sporting alternative to killing members of a depleted population, the emphasis such anglers place on quiet contemplation fits awkwardly with their advice on how to capture prey. There is a basic irony in the notion that the angler derives a sense of peaceful belonging from engaging in a fundamentally violent recreation. By comparison, the frank presentation of killing in early modern texts requires readers to imagine the act of taking a life as the central joy of the angler. The author of the *Treatyse* promises that it "shall be to you a very pleasure to se the fayr bright shynynge scalyd fysshes dysceyued [deceived] by your crafty meanes and drawen upon londe."[13] Other writers make liberal use of the verb *to kill*. Dennys's speaker, for example, is not shy about the ultimate purpose of the aquatic hunt: after procuring the right bait, "Then shall you haue them byte as you would wish: / And ready sport to take your pleasure still, / Of any sort that best you like to kill."[14] In a short advisory poem by "B. R." which is printed at the end of *The Secrets of Angling*, the easy rhyming of that crucial verb is again on display: "Fish where thou wilt, thou shalt have sport thy fill, / When twenty faile, thou shalt be sure to kill."[15] And in Thomas Barker's treatise *The Art of Angling*, *kill* and *pleasure* are both frequently used to describe the angler's motivation, sometimes together: "The naturall Flie is sure Angling, and will kill great store of Trouts with much pleasure."[16]

The point is clearly to catch fish, draw them out of their habitats, and end their lives. But killing isn't the whole story; as Garry Marvin argues:

> The primary interest of the sports hunter is not that of obtaining meat nor even that of merely killing an animal. Rather, it is with an

> immersion into the very difficulty of bringing about an encounter with the animal and with the pleasure and satisfaction that comes from successfully overcoming these self-imposed restrictions and difficulties.[17]

Hunting—or "wild killing"—is distinct from the many forms of "domestic killing" identified by Marvin in that "the death of the animal has to be won from the animal rather than its life simply being taken."[18] The feeling of taking up such a challenge is voiced repeatedly in the angling literature of early modernity, where encounters with fish result from much careful study and thankless practice.

In accepting a rhetorical distinction between fishing and other kinds of hunting (a distinction that goes back at least as far as the medieval period), contemporary treatments of the sport have tended to focus less on the suffering and death of individual animals, and more on the ways in which anglers have contributed to the conservation of habitats and species. Given that angling may not, on balance, have proved as beneficial to ecosystems as we might have hoped, it might now be time to look differently at a sport that has long been touted for the "harmlesse pleasure" it brings.[19] For a start, to put killing back into the balance is to accept that fishing is a hunt—in some ways different from the pursuit of deer or rabbits, but in many ways dependent for its cultural meaning upon the practice of capturing mammals in a relative wilderness. At its emergence into English literature, the idea of angling was inflected by the social signification of hunting. As Jeffrey Theis has shown, "The history of hunting tells us that poaching is rarely about finding dinner; rather, poaching is enmeshed in social privilege and control of the English landscape."[20] As hunting became a protected aristocratic sport over the course of the medieval and early modern periods, the hunting practices of ordinary people were criminalized.[21] This shift is reflected in the effort to police angling and to limit people's access to knowledge about the sport.

The protectiveness surrounding sportfishing also contributes to its figuration as the quintessential recreation, unmoored from the claims of growling stomachs. Angling is pure leisure—for Antony and Cleopatra, it marks an utter disregard for the labor of leadership and the hardship of battle; for the fishers in the *Urania*, it is so totally optional that it can be abandoned at a moment's notice and replaced with some other activity. The writers of fishing treatises sometimes describe fish as food, but they give much more time to what goes on between humans and their potential prey at the riverside. Even the birds who gobble up fry in Donne's *Metempsychosis* do so on a whim, motivated by overweening greed rather than genuine hunger. In Walton's *Compleat Angler*, men have an appetite for companionship and good humor above all; they may eat a fish or two, but they are more likely to give them away to those who are in need of sustenance. Like deer-poaching, angling was "rarely about finding dinner," and had much more to do with the performance of social roles and relations.

It remains true, however, that the literature of angling that emerged in early modern England was a dynamic repository of detailed information about the lives of fishes. Without an English ichthyology at hand, writers of fishing treatises took up the challenge of representing their sport as a vernacular recreation, tied to the geography of English rivers and the nature of English people. The result is a plump record of the life histories of freshwater fish that attests to the biodiversity of early modern waterways and the reach of human curiosity about alien creatures. The killing at the heart of these projects should not be ignored, but there's more to see. There are traces of slime on the pages that document an early modern poetics of angling—places where human bodies and minds were touched, physically and figuratively, by fish. The residue of these haptic moments lingers. In Donna Haraway's formulation, "touch ramifies and shapes accountability." She says, "In touch and regard, partners willy nilly are in the miscegenous mud that infuses our bodies with all that brought that contact into being. Touch and regard have consequences."[22] Facing fish, getting beslimed by them, letting them go—these are all methods of learning that we can draw up from the angling literature of early modernity and apply to new questions about how we come to know and care for other beings. Julian Yates, laying out his concept of "multispecies impressions," explains that our attention to the touches we share with other life forms need not be construed as inescapably inward-looking: "Following Haraway, the positing of a multispecies does not represent an always already compromised anthropocentric endeavor but … an open system or inquiry, the posing of a set of questions to the various beings we encounter as to what counts for them."[23] If, in keeping with the curiosity of early modern angler-authors, we pause to ask what counts for fish, we might find that we have shared interests after all.

## Notes

1 Marjorie Swann, introduction to *The Compleat Angler* by Izaak Walton and Charles Cotton (Oxford: Oxford University Press, 2014), xx–xxi.
2 A. Dionys de Leeuw, "Can Nature Conservation Justify SportFishing?" *Environmental Ethics* 34 (2012).
3 John McDonald, *The Origins of Angling*, 66.
4 de Leeuw, "Can Nature," 165.
5 de Leeuw, "Can Nature," 174.
6 James Prosek, "Trout Fishing, a Taut Line to Our Past," *New York Times*, April 18, 2015.
7 [William Samuel], *The Arte of Angling* (1577), ed. Gerald Eades Bentley (Princeton, NJ: Princeton University Press, 1956), 32.
8 [John Dennys], *The Secrets of Angling* (1613, STC [2nd ed.] 6611), D5r.

9 Dennys, *The Secrets*, D5r.
10 Dennys, *The Secrets*, D3r.
11 Zach Matthews, "October Browns," *Kentucky Afield*, Fall 2009, 9.
12 Matthews, "October Browns," 11.
13 *Treatyse of fysshynge wyth an Angle*, in *This present boke shewyth the manere of of hawkynge [and] huntynge and also of diuysynge of cote armours ... [The Book of Saint Albans*], (1496, STC [2nd ed.] 3309), i4r.
14 Dennys, *The Secrets*, D1r.
15 Dennys, *The Secrets*, E3v.
16 Thomas Barker, *The Art of Angling* (1653, Wing B783), 6.
17 Garry Marvin, "Wild Killing: Contesting the Animal in Hunting," in *Killing Animals*, by The Animal Studies Group (Urbana: University of Illinois Press, 2006), 19.
18 Marvin, "Wild Killing," 12.
19 Dennys, *The Secrets*, B1r.
20 Jeffrey Theis, "The 'ill kill'd' Deer: Poaching and Social Order in *The Merry Wives of Windsor*," *Texas Studies in Literature and Language* 43, no. 1 (2001): 47.
21 Theis, "The 'ill kill'd' Deer," 47–48.
22 Donna J. Haraway, *When Species Meet* (Minneapolis: University of Minnesota Press, 2007), 36.
23 Julian Yates, *Of Sheep, Oranges, and Yeast: A Multispecies Impression* (Minneapolis: University of Minnesota Press, 2017), 21.

## References

Barker, Thomas. *The Art of Angling*. London, 1653. Wing B783.

de Leeuw, A. Dionys. "Can Nature Conservation Justify SportFishing?" *Environmental Ethics* 34, no. 2 (2012): 159–75.

[Dennys, John]. *The Secrets of Angling*. London, 1613. STC (2nd ed.) 6611.

Donne, John. *The Complete Poems of John Donne*, edited by Robin Robbins. Harlow, UK: Pearson Education, 2010.

Haraway, Donna J. *When Species Meet*. Minneapolis: University of Minnesota Press, 2007.

McDonald, John. *The Origins of Angling*. New York: Doubleday, 1957.

Marvin, Garry. "Wild Killing: Contesting the Animal in Hunting." In *Killing Animals*, by The Animal Studies Group, 10–29. Urbana: University of Illinois Press, 2006.

Matthews, Zach. "October Browns." *Kentucky Afield*, Fall 2009.

Prosek, James. "Trout Fishing, a Taut Line to Our Past," *New York Times*, April 18, 2015.

Samuel, William. *The Arte of Angling*. 1577. Edited by Gerald Eades Bentley. Princeton, NJ: Princeton University Press, 1956.

Swann, Marjorie, ed. Introduction to *The Compleat Angler*, by Izaak Walton and Charles Cotton. Oxford: Oxford University Press, 2014.

Theis, Jeffrey. "The 'ill kill'd' Deer: Poaching and Social Order in *The Merry Wives of Windsor*." *Texas Studies in Literature and Language* 43, no. 1 (2001): 46–73.

*This present boke shewyth the manere of of hawkynge [and] huntynge and also of diuysynge of cote armours* … [The Book of Saint Albans]. Westminster, 1496. STC (2nd ed.) 3309.

Yates, Julian. *Of Sheep, Oranges, and Yeast: A Multispecies Impression*. Minneapolis: University of Minnesota Press, 2017.

Walton, Izaak. *The Compleat Angler or, the Contemplative Man's Recreation*. (1st ed). London, 1653. Wing (2nd ed.) W661.

# Index


"The admonition by the Auctor, to all yong Gentilwomen" 1–2, 16–17, 143
*The Aeneid* 142
agroecology 28
agro-ecosystems 28
Alaimo, Stacy 36–39
Amyot, Jacques 96; *Les Vies des hommes illustres, grecs et romains* 96–99, 104, 106, 108, 110
anadromy 13–14, 64–66
angle (hook) 102
angling: appropriate attire for 12, 78; compared with other sports 36, 99–100; as contemplative recreation 89, 118–19; as craft 27; as deception 103, 117; gendering of 143, 145; as hunting 104, 127, 161; as social activity 140; spiritual benefits of 36–37
*Antony and Cleopatra* 96–112, 130
*Arcadia* 117–18
*The Art of Angling* (1651), 109, 144–45
*The Arte of Angling* (1577) 101, 122, 140, 158
Athenagoras 57–58

Bacon, Francis 139
bait 39–43
"The Bait" 50–51, 103
barb, on fish hook 102–3
Barker, Thomas 109, 144–45
Bellamy, Elizabeth J. 13
Belon, Pierre 6–7, 8–9, 7
Berners, Juliana 24, 44n2
*Book of Saint Albans* 24–26
*A Booke of fishing with Hooke & Line* 66–**67**, 101–2, 109
Brayton, Dan 11–12
brown 11–12
Buell, Lawrence 15, 29, 34, 82

catadromy 13–14
catch-and-release 160
caterpillars 146
*Certaine experiments concerning fish and fruite* 51–53
chub 140–41, 144
cock-fighting 99–100
cognition, distributed 4–5
cognition, in fish 4
cognitive ecology 14–17, 51, 70n10, 76, 82, 145, 148–49; *see also* learning; plasticity; signal detection theory
commensal species 40–41
*The Compleat Angler* 10, 49–50, 134–51
conceptual metaphor theory 15–16, 81–82, 156
conservation 156–60
conversion to angling 136, 145–46
cooking fish 144–45; *see also* recipe
cork *see* fishing tackle

*De aquatilibus* 6–7, 8–9
de Leeuw, Dionys 157–58
Dennys, John 12, 75–92, 103, 109, 120, 158–59
de Worde, Wynkyn 24–26, 44
disgust 42–44
Donne, John 49–69; conversion 150; education 149–50; sermons 51, 57–60; *see also* "The Bait;" "An Elegy upon the Death of Mistress Bulstrode;" *Metempsychosis*; "Sir, more than kisses;" "A Valediction Forbidding Mourning"
Du Bartas, Guillaume 146

*Eclogues* 142, 144
ecology (of angling) 14–16, 29, 31–32, 35–38; *see also* cognitive ecology
ecosystems 2, 16, 28–29, 32, 37–41, 78, 82, 88, 119, 156–58
eel 13, 53, 102, 147
"An Elegy upon the Death of Mistress Bulstrode" 51, 62
empiricism 139
entomophagy 40–42
Erasmus, Desiderius 62
erosion 111–12

fish: drawn up by anglers 106–10, **106–8**; eggs *see* roe; as food 39, 144–5; fry 60–1; larvae 55–6; metamorphosis 55; reproduction 51–4; respiration 63–4; silence of 61–4; species 89–90; *see also* barbell; chub; eel; minnow; pike; salmon, stickleback; trout
fishing tackle 122; cork 124; flies 31; hooks 31–2, **32**, 102–3, 124; line 30, 80–81; net 109, 130–31; rod 29–31, 123–24
flag (plant) 101–2
Fletcher, Phineas 143
flies (artificial) *see* fishing tackle
flies (insects) 147
freshwater systems 3, 9, 11–13, 86, *see also* ponds; rivers
fresh water vs. salt water 11–12, 13, 65, 97
fruit flies 149
Fudge, Erica 3–4

Gessner, Konrad 5–10, **7**
Graham, Elspeth 3, 10, 12
Grubb, Thomas 16–17

Haraway, Donna 162
harm to fish 158
*Henry V* 109
Herbert, William 116
*Historiae animalium* 5–6, **7**
*Historie of Foure-footed Beastes* 10–11
home ecologies 21n55, 29–32
*The Honest Whore* 143
hooks *see* fishing tackle
Hopkins, Gerard Manley 82
hunting: history of 161

indirect development 55, 146
insects: and disgust 42; larvae 146–47; as living property 42; edible 40–42; *see also* caterpillars, flies, fruit flies, maggots
interactionism 15
Izaak Walton League 156–57

Jonson, Ben 116–17
Justin Martyr 57–58

Kilgour, Maggie 13
killing 160–61

larval diapause 148
lateral line 4, 33
learning 27, 76, 88–92, 139–41, 148–50
leisure 25, 34, 97–106, 144, 161
*Libri di piscibus marinis* **6–7**, 8–9
*Life of Donne* 149–50
line *see* fishing tackle
localism 27–28
love: of angling 141, 146; between anglers 136–37

maggots 41, 147
Markham, Gervase 100
Martin, Randall 27–28
Mascall, Leonard 66–**67**, 101–2, 109
Matthews, Zach 159–60
Mentz, Steve 11–12
metamorphosis 55, 146–49
*metanoia* 150
metaphor 123–24
*Metempsychosis* 51–57, 60–69
migration (of fish) 13–14; *see also* anadromy
minilivestock 41
minnow 145
Murray, Molly 150–51

neighbor-love 43
net *see* fishing tackle
North, Thomas 96; *Lives of the Noble Grecians and Romanes* 96–100, 104–6, 108, 110–11
Nussbaum, Martha 42–43

Offley, John 137
osiers 119–20, 125
Ovid 83, 87–88
oystercatcher 66–68

pastoral mode 142
Penshurst 116–17
people-watching 98–99
*Pericles* 109
pike 62–3, 90, 109, 139–40
*Piscatorie Eclogs* 143
plasticity 5, 91, 145
Plutarch 96; *Lives* 96–100, 104, 110, 112
ponds 11, 18–19n6, 51–52, 56, 117, 129–30
Prosek, James 158

rain 12, 78, 141–43, 148
recipe 139–40, 145
resurrection, doctrine of 57–9
riparian zones 17, 76, 88, 97, 101, 111–12, 117, 119, 149, 160
river poetry, Tudor 84–5
rivers 9, 11, 13, 32, 34, 50, 52, 75–76, 83–86, 97, 162; flooding of 111; as sources of information 86; as symbols 85; *see also* riparian zones
rod *see* fishing tackle
roe 51–6
*Romeo and Juliet* 120
Rondelet, Guillaume **6–7**, 8–9

salmon 7, 13–14, 65, 143
salt-fish 104
salt water *see* fresh water vs. salt water
Samuel, William 101, 122, 140, 158
sea-pie *see* oystercatcher
*The Secrets of Angling* 75–92, 103, 109, 120, 124, 158–59; use of classical sources in 82–84
*La Sepmaine* 146
Shakespeare, William 96–112; *see also Antony and Cleopatra*; *Henry V*; *Pericles*; *Romeo and Juliet*; *The Winter's Tale*
Sidney family 116
Sidney, Philip 117
signal detection theory 16–17, 82
"Sir, more than kisses" 61, 63
skill 79–80
slime 12, 101–2, 162
smolt 65
song 121–22, 136, 143, 151
spontaneous generation 53, 147
sport 97–100, 106
stickleback ("sticklebag") 145
Swann, Marjorie 8–9, 136, 140, 142, 156–57

tackle *see* fishing tackle
Taverner, John 51–53
teaching 87–88, 137, 148
Theis, Jeffrey 161
Topsell, Edward 10–11, 146
trans-corporeality 36–39
*Treatyse of fysshynge wyth an Angle* 14, 24–44, **32**, **33**, 103
Tribble, Evelyn 4–5
trout 7, 10, 12, 134–35, 159
Tuana, Nancy 15

*Urania* 116–31

"A Valediction Forbidding Mourning" 63–64
Virgil 142

Walton, Izaak 10, 49, 134–51
whale 69
Whitney, Isabella 1–2, 16–17, 143
*The Winter's Tale* 110, 113–14n32
women anglers 101–5, 117–27, 131, 143, 145
Woolf, Virginia 80–81
Worms 59
Wotton, Henry 49–50, 61
Wroth, Mary 116–31

Yates, Julian 162

For Product Safety Concerns and Information please contact our EU
representative GPSR@taylorandfrancis.com
Taylor & Francis Verlag GmbH, Kaufingerstraße 24, 80331 München, Germany

www.ingramcontent.com/pod-product-compliance
Ingram Content Group UK Ltd.
Pitfield, Milton Keynes, MK11 3LW, UK
UKHW021300190926
475701UK00009B/90
*9780367664930*